Ferdinand Mueller

Iconography of Australian Salsolaceous Plants

Ferdinand Mueller

Iconography of Australian Salsolaceous Plants

ISBN/EAN: 9783337319618

Printed in Europe, USA, Canada, Australia, Japan

Cover: Foto ©berggeist007 / pixelio.de

More available books at **www.hansebooks.com**

ICONOGRAPHY

OF

AUSTRALIAN SALSOLACEOUS PLANTS.

BY

BARON FERD. VON MUELLER, K.C.M.G., M. & PH.D., F.R.S.,

GOVERNMENT BOTANIST OF THE COLONY OF VICTORIA.

"Multi enim sunt vocati, pauci vero electi."—*Evang. Matth.* xxii, 14.

FIRST DECADE.

By Authority:
ROBERT S. BRAIN, GOVERNMENT PRINTER, MELBOURNE.
1889.

TO

HIS EXCELLENCY

THE RIGHT HONORABLE THE EARL OF HOPETOUN,

G.C.M.G.,

Governor of the Colony of Victoria,

THIS VOLUME,

ELABORATED MAINLY FOR ADVANCING SOME OF THE PASTORAL INTERESTS

IN

HER MAJESTY'S AUSTRALIAN DOMINIONS,

Is most Reverrally Dedicated.

In a desire of continuing the illustrative volumes, issued already on Eucalyptus, Acacia and allied genera, as well as on Myoporinous plants, it was deemed best, to select the Salsolaceae for next sequence, inasmuch as this ordinal group in the vegetation of Australia presents not only a multitude of endemic forms of high phytologic interest, but also a considerable number of prominent utilitarian value. The "Saltbushes" constitute in many wide tracts of our island-continent the prevailing vegetation, and on this again depends then locally to a large extent the sustenance of herds and flocks. Moreover, this kind of pastoral nutriment has proved so particularly wholesome, that "Saltbush-country" has become among Australian ruralists quite famous already for a long series of years, more particularly so as Salsolaceae will live even through the direst periodic droughts. That under such circumstances these important plants may finally succumb through continuous depasturing processes, cannot be surprising; thus, then the necessity is forced on the proprietors or holders of "runs," to renew the saltbush-vegetation by methodical sowings. Furthermore many pastoral estates could, on adequate soil, doubtless be largely improved by the introduction of the best kinds of these plants as additions to existing natural herbage. It becomes then imperative also, to select only those particular species, which are preferentially liked by pasture-animals, for any portion of this part of the world, to resuscitate failing nutriture. Yet such selections would often be of the utmost difficulty, unless from pictural displays in a special work each of the numerous kinds of our salsolaceous herbs or shrubs could be readily recognised. But, irrespective of our own motives for practical gain, we here should remain conscious, that while we are constantly adding from abroad to the plants-treasures of Australia, we likewise in a cosmopolitan spirit should afford facilities in return, to select from the Australian gifts of nature whatever might be conducive for increasing also the riches of rural pursuits in any other part of the world, with a genuine and disinterested desire for adding thus from here to the comfort and prosperity also in many another land through circumspect benignity and due gratefulness of ours.

Melbourne, October, 1889

ATRIPLEX FISSIVALVE.

F. v. Mueller, fragmenta phytographiae Australiae ix. 123 (1875).

PLATE I.

1, cluster of unexpanded staminate flowers.

2 and 3, staminate flowers.

4, front- and back-view of a stamen.

5, pollen-grain.

6 and 7, pistillate flowers.

8, a fruit, one half of the calyx removed.

9, longitudinal section of a fruit with half of its calyx.

10, a seed.

11, transverse section of a fruit.

12, longitudinal section of a fruit.

All enlarged, but to various extent.

I

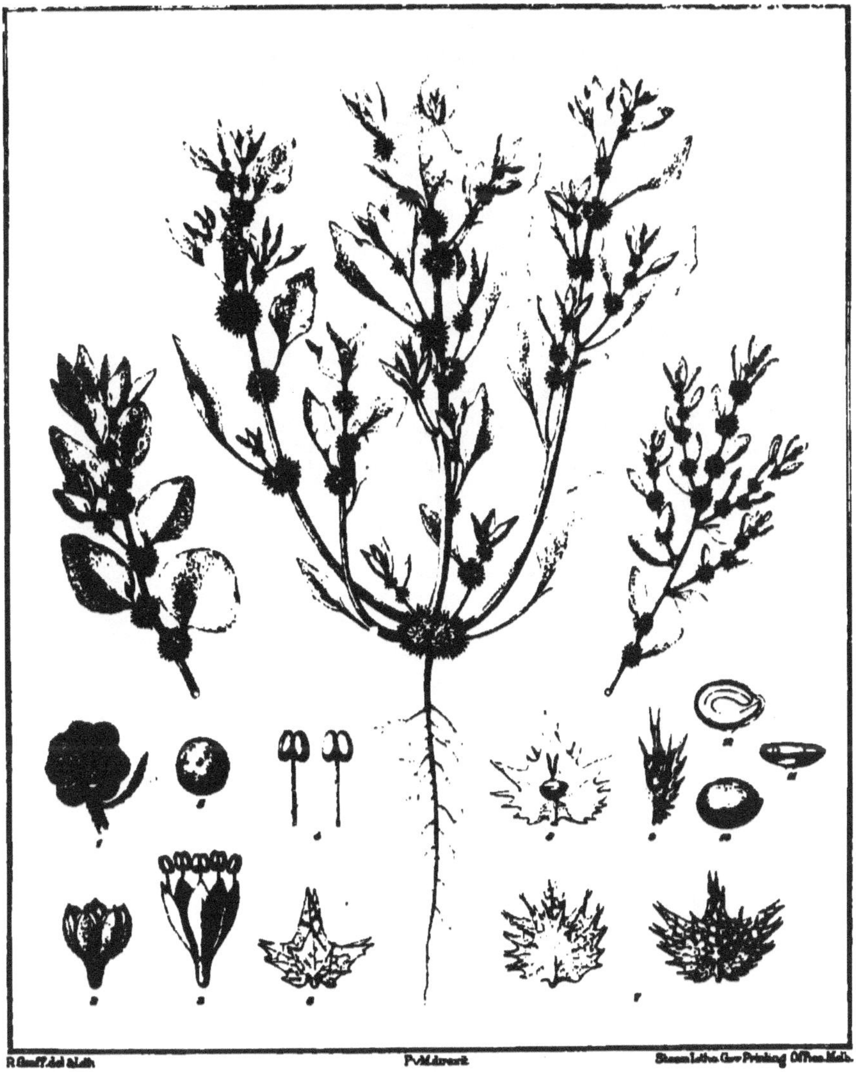

R Graff del & Lith F v M direxit Steam Litho Gov Printing Office Melb.

Atriplex fissivalve *F.v.M.*

ATRIPLEX CRYSTALLINUM.

J Hooker in London Journal of Botany vi. 279 (1847).

PLATE II.

1, leaves.
2, portion of a branchlet with a cluster of flowers.
3, a staminate flower.
4, back- and front-view of a stamen.
5, pollen-grain.
6, a pistillate flower.
7, a pistillate flower, half of the calyx removed.
8, a fruit-bearing calyx.
9, a fruit, half of its calyx removed.
10, longitudinal section of a fruit with half its calyx.
11, a seed.
12, transverse section of a fruit.
13, longitudinal section of a fruit.

All enlarged, but to various extent.

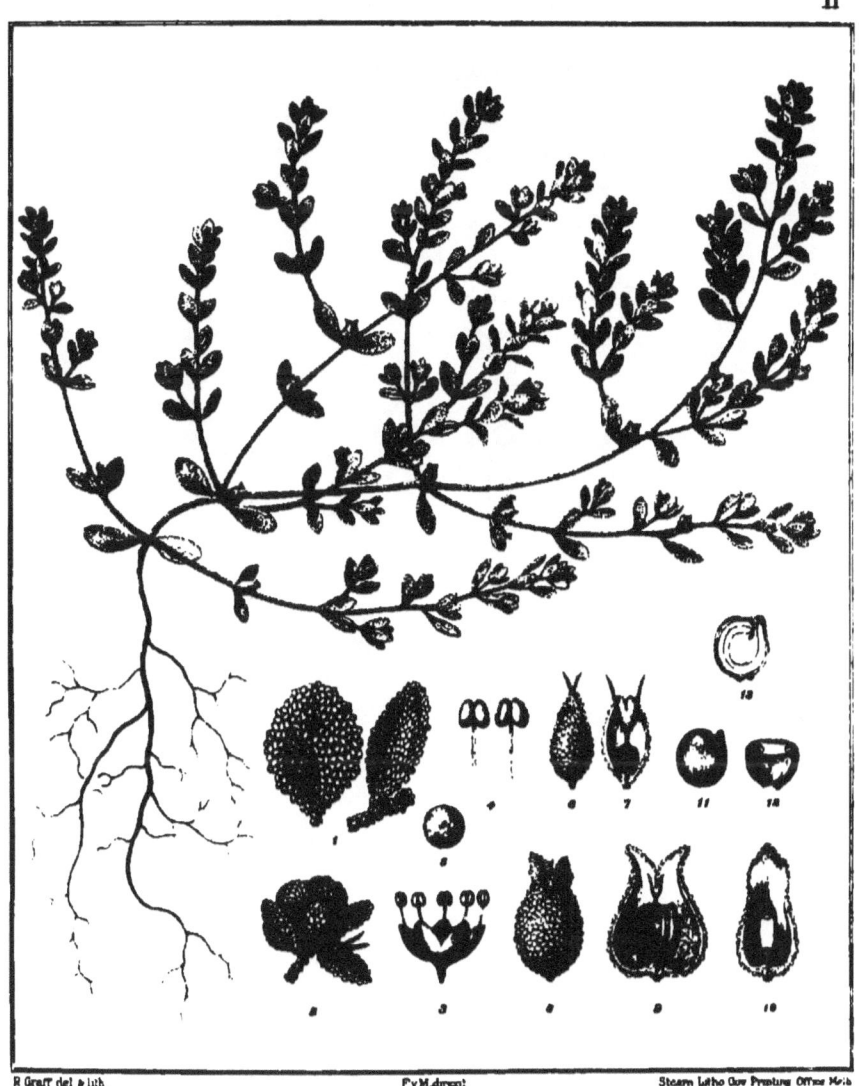

Atriplex crystallinum J.Hooker.

ATRIPLEX LEPTOCARPUM.

F. v. M. in Transactions of the Philosophical Institute of Victoria ii. 74 (1857).

PLATE III.

1, a cluster of staminate and of pistillate flowers.

2, a staminate flower.

3, front- and back-view of a stamen.

4, pollen-grain.

5, pistillate flowers.

6, longitudinal section of a fruit with half of its calyx.

7, a fruit.

8, a seed.

9, transverse section of a seed.

10, longitudinal section of a seed.

All enlarged, but to various extent.

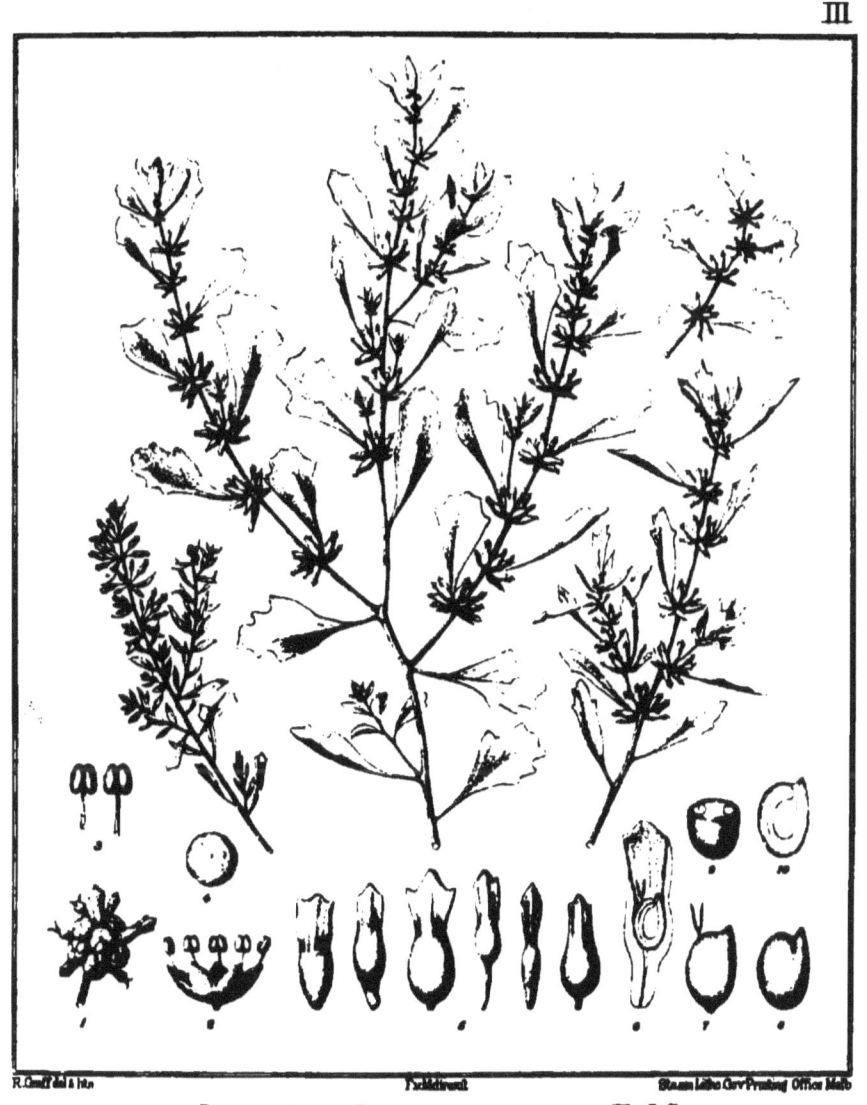

Atriplex leptocarpum F.v.M.

Atriplex limbatum.

Bentham, Flora Australiensis v. 178 (1870).

PLATE IV.

1, cluster of staminate and of pistillate flowers.

2, staminate flower.

3, front- and back-view of a stamen.

4, pollen-grain.

5 and 6, pistillate flowers.

7, a fruit, one half of the calyx removed.

8, transverse section of a seed.

9, longitudinal section of a fruit.

All enlarged, but to various extent.

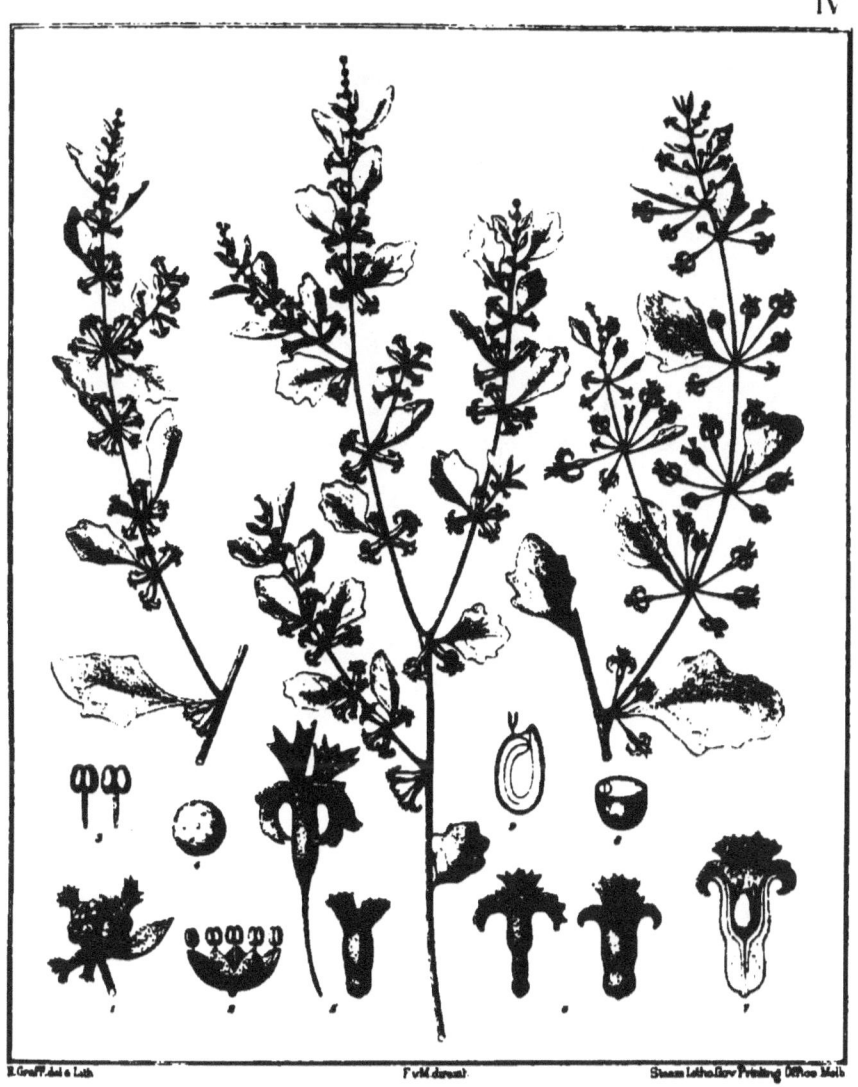

Atriplex limbatum Bentham.

ATRIPLEX VELUTINELLUM.

F. v. Mueller, Report on Plants of Babbage's Expedition 20 (1858).

PLATE V.

1, a staminate flower.

2, front- and back-view of a stamen.

3, pollen-grain.

4, three pistillate flowers.

5, two fruits, one half of the calyx removed.

6, longitudinal section of a fruit with half of its calyx.

7, a seed.

8, longitudinal section of a fruit.

All enlarged, but to various extent.

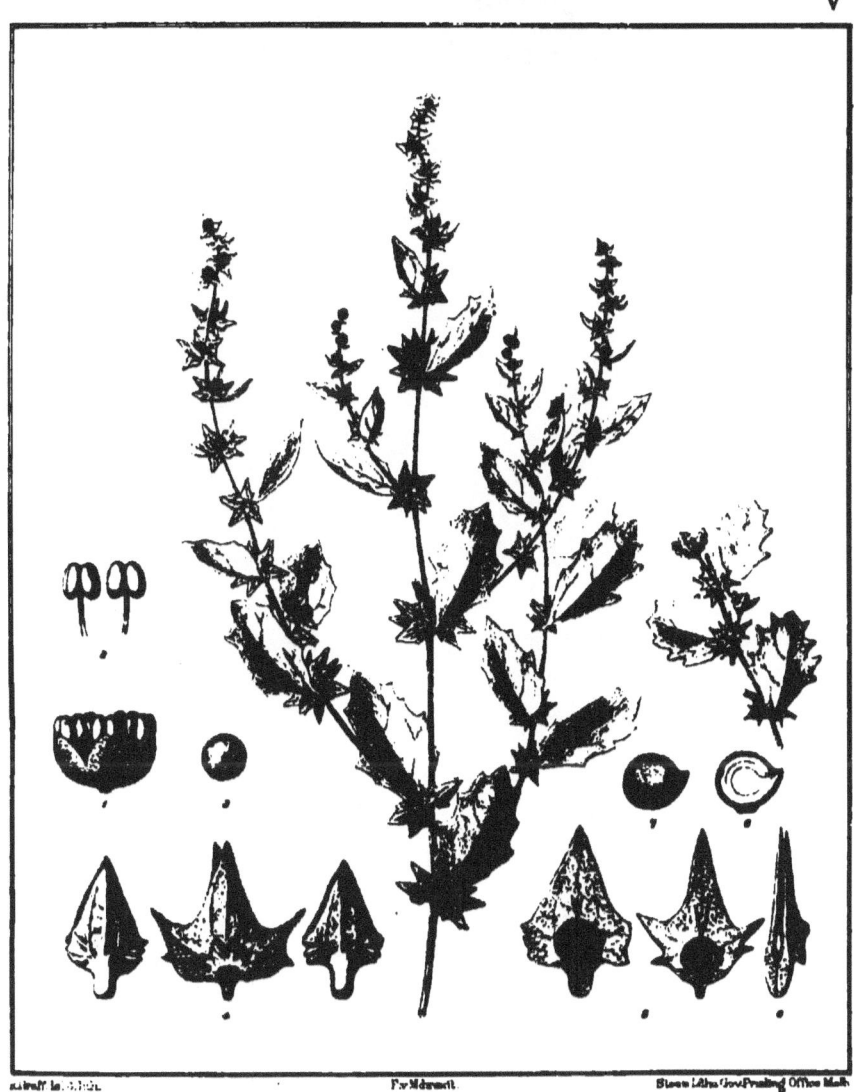

Atriplex velutinellum F.v.M

ATRIPLEX LOBATIVALVE.

F. v. Mueller, inedited.

PLATE VI.

1, branchlet with staminate and pistillate flowers.

2, a staminate flower.

3, front- and back-view of a stamen.

4, pollen-grain.

5, three pistillate flowers.

6, a fruit, half of its calyx removed.

7, longitudinal section of a fruit with half of its calyx.

8, a seed.

9, transverse section of a fruit.

10, longitudinal section of a fruit.

All enlarged, but to various extent.

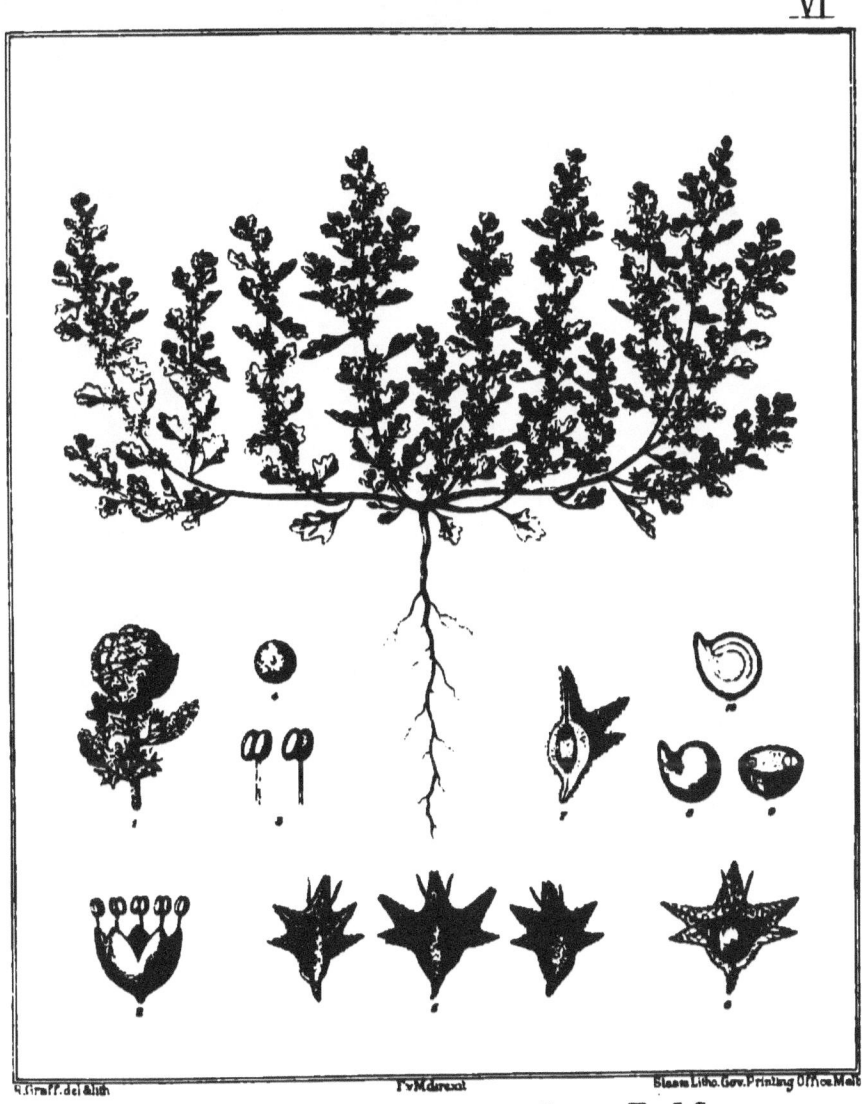

Atriplex lobativalve *F.v.M.*

Atriplex Muelleri.

Bentham, Flora Australiensis v. 175 (1870).

PLATE VII.

1, portion of a branchlet with a cluster of unexpanded flowers.

2, a staminate flower, expanded.

3, front- and back-view of a stamen.

4, pollen-grain.

5, five pistillate flowers.

6, a fruit, half of its calyx removed.

7, longitudinal section of a fruit with half of its calyx.

8, a seed.

9, transverse section of a fruit.

10, longitudinal section of a fruit.

All enlarged, but to various extent.

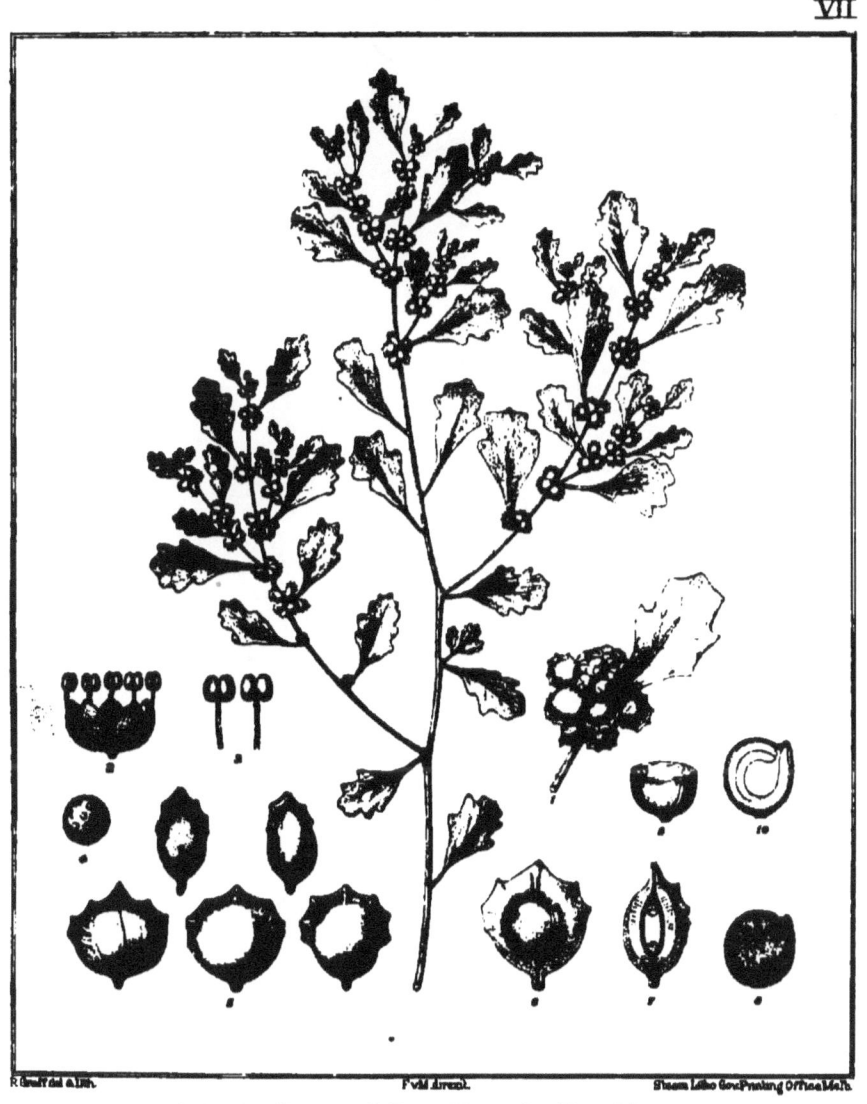

Atriplex Muelleri *Bentham.*

ATRIPLEX SEMIBACCATUM.

R. Brown, prodromus florae Novae Hollandiae 406 (1810).

PLATE VIII.

1, cluster of unexpanded flowers.

2, expanded staminate flower.

3, front- and back-view of a stamen.

4, pollen-grain.

5, six pistillate flowers.

6, a pistillate flower, one half of the calyx removed.

7 and 8, longitudinal sections of fruit, half of the calyx removed.

9, a seed.

All enlarged, but to various extent.

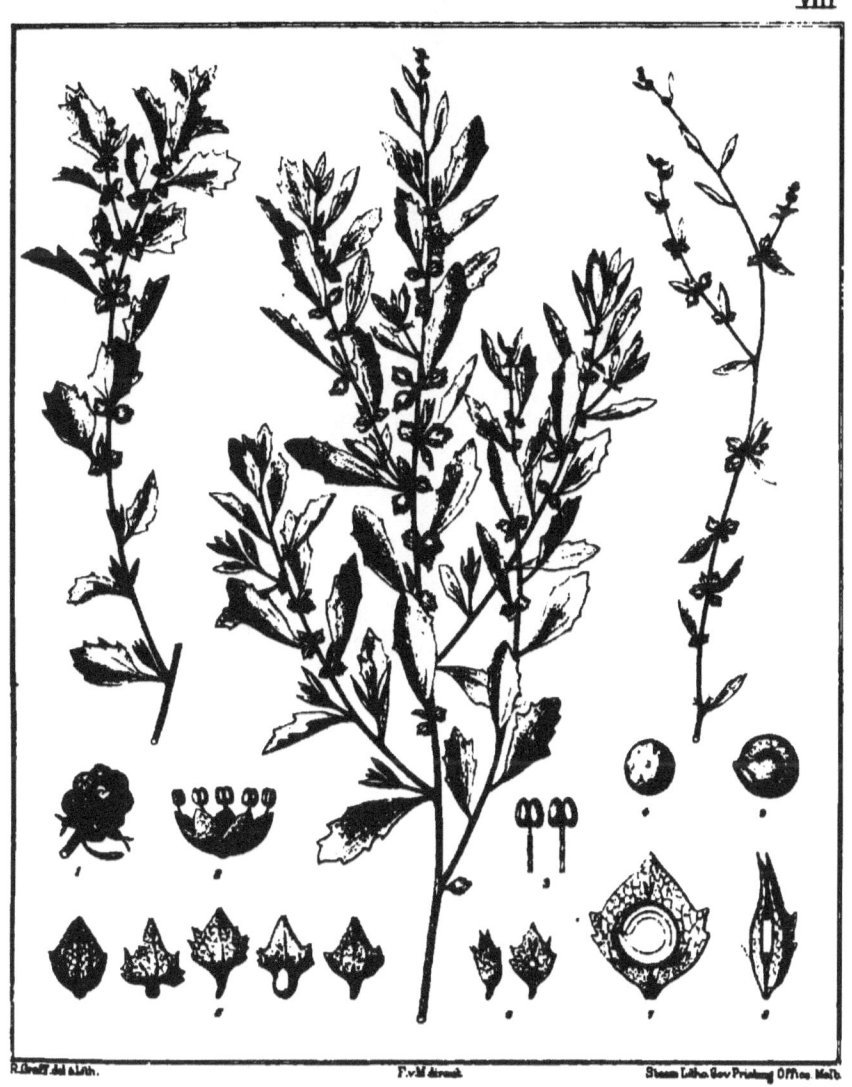

Atriplex semibaccatum R.Brown.

ATRIPLEX HUMILE.

F. v. Mueller, fragmenta phytographiae Australiae iv. 48 (1866).

PLATE IX.

1, portion of a branchlet with unexpanded flowers.
2, expanded staminate flower.
3, front- and back-view of a stamen.
4, pollen-grain.
5, two pistillate flowers.
6, three fruit-bearing calyces.
7, a fruit-bearing calyx, half of it removed.
8, longitudinal section of a fruit-bearing calyx.
9, a seed.
10, transverse section of a seed.
11, longitudinal section of a seed.

All enlarged, but to various extent.

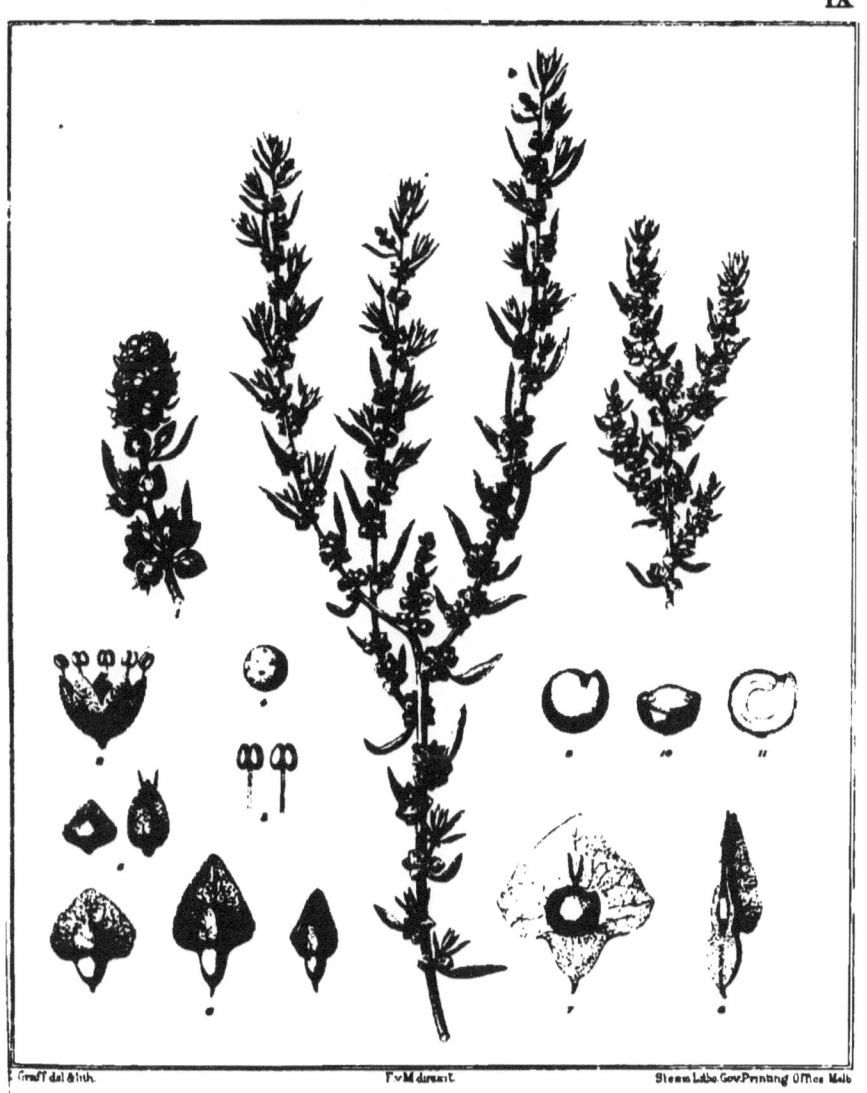

Atriplex humile F.v.M

ATRIPLEX PROSTRATUM.

R. Brown, prodromus florae Novae Hollandiae 406 (1810).

PLATE X.

1, portion of a branchlet with unexpanded flowers.

2, a staminate flower.

3, front- and back-view of a stamen.

4, pollen-grain.

5 and 6, five pistillate flowers.

7, a fruit, half of its calyx removed.

8, longitudinal section of a fruit, half its calyx removed.

9, transverse section of a seed.

10, longitudinal section of a fruit.

All enlarged, but to various extent.

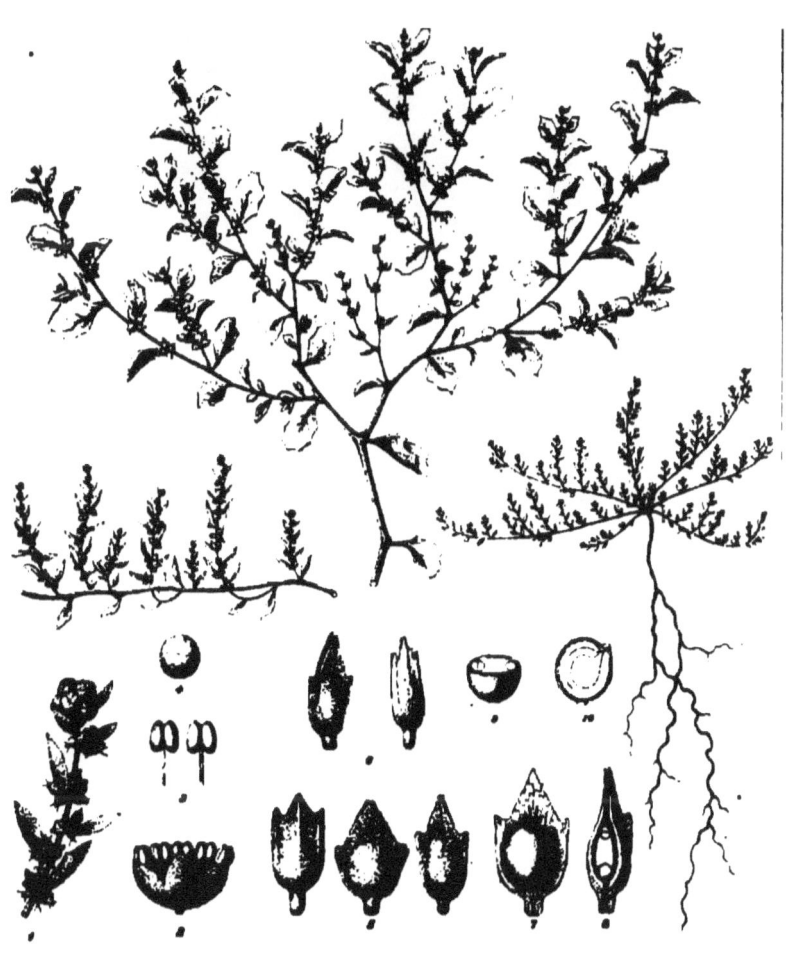

Atriplex prostratum R.Brown.

ICONOGRAPHY

OF

AUSTRALIAN SALSOLACEOUS PLANTS.

BY

BARON FERD. VON MUELLER, K.C.M.G., M. & PH.D., F.R.S.,

GOVERNMENT BOTANIST OF THE COLONY OF VICTORIA.

"Multi enim sunt vocati, pauci vero electi."—*Evang. Matth. xxii, 14.*

SECOND DECADE.

By Authority:
ROBERT S. BRAIN, GOVERNMENT PRINTER, MELBOURNE.
1890.

Atriplex angulatum.

Bentham, Flora Australiensis v, 174 (1870).

PLATE XI

1, cluster of staminate and of pistillate flowers.
2, a staminate flower.
3, front- and back-view of a stamen.
4, pollen-grain.
5, five pistillate flowers.
6, a fruit, one half of the calyx removed.
7, an exceptional pistillate flower with one stamen.
8, twin-fruits within one calyx, half of the latter removed.
9, longitudinal section of a fruit with half its calyx.
10, a fruit, separated.
11, a seed.
12, transverse section of a fruit.
13, longitudinal section of a fruit.

All enlarged, but to various extent.

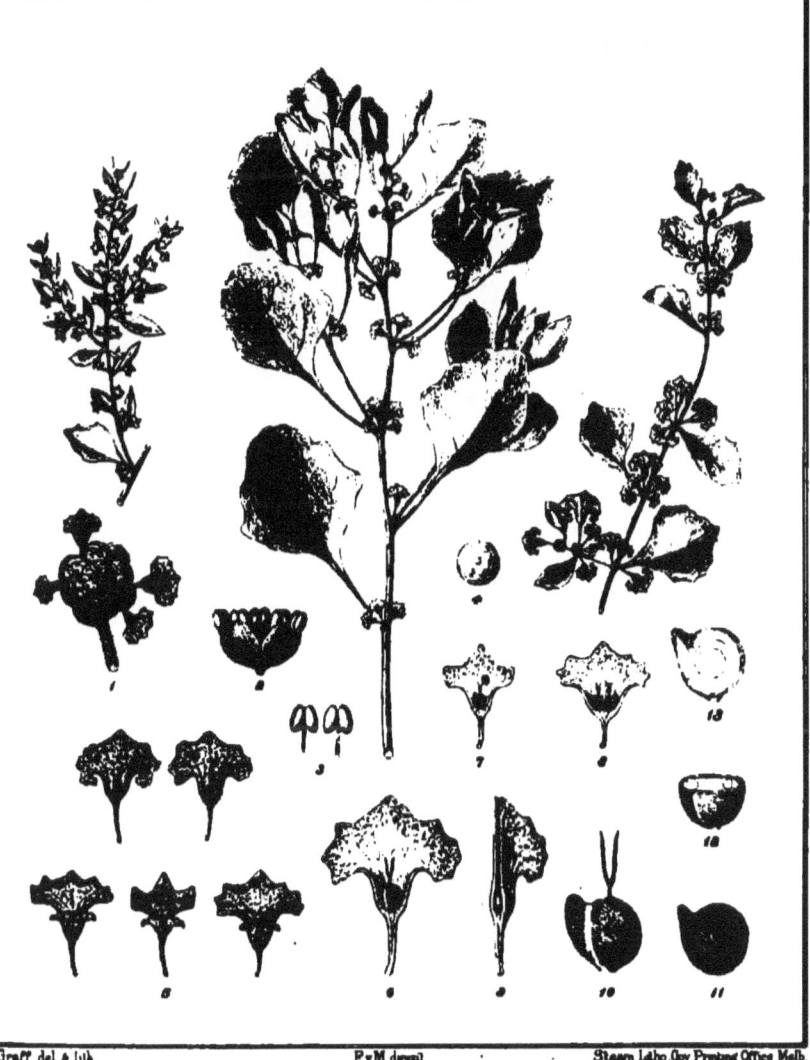

Atriplex angulatum Bentham.

Atriplex quinii.

F. v. Mueller in Victorian Naturalist v, 96 (1888).

PLATE XII.

1, separate leaves, one fragmentary.
2, three pistillate flowers.
3, a staminate flower.
4, front- and back-view of a stamen
5, pollen-grain.
6, three fruit-bearing calyces.
7, a fruit-bearing calyx, one half of it removed.
8, longitudinal section of a fruit-bearing calyx.
9, a seed.
10, transverse section of a seed.
11, longitudinal section of a seed.

All enlarged, but to various extent.

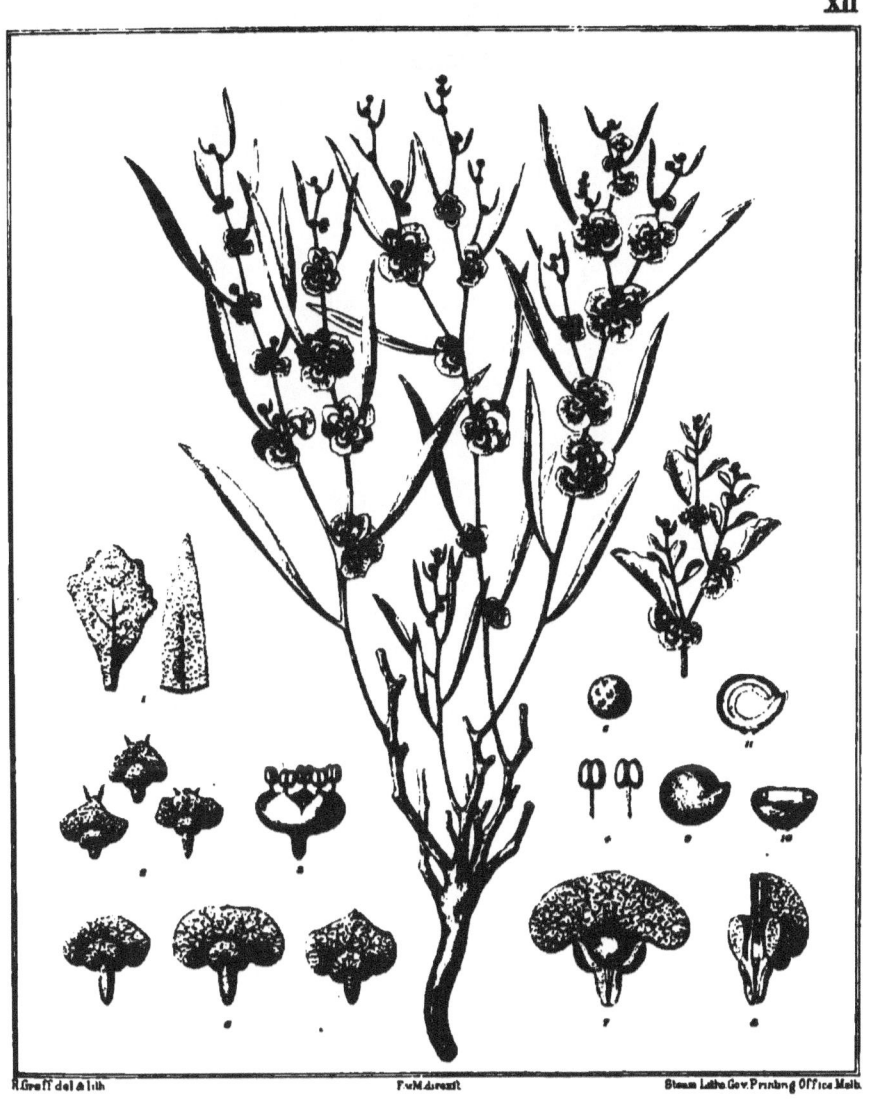

Atriplex Quinii F.v.M.

ATRIPLEX STIPITATUM.

Bentham, Flora Australiensis v, 168 (1870).

PLATE XIII.

1, portion of two leaves.

2, cluster of unexpanded staminate flowers.

3, expanded staminate flower.

4, front- and back-view of a stamen.

5, pollen-grain.

6, cluster of pistillate flowers.

7 and 8, longitudinal sections of fruit, half the calyx removed.

9, a seed.

10, longitudinal section of a fruit.

All enlarged, but to various extent.

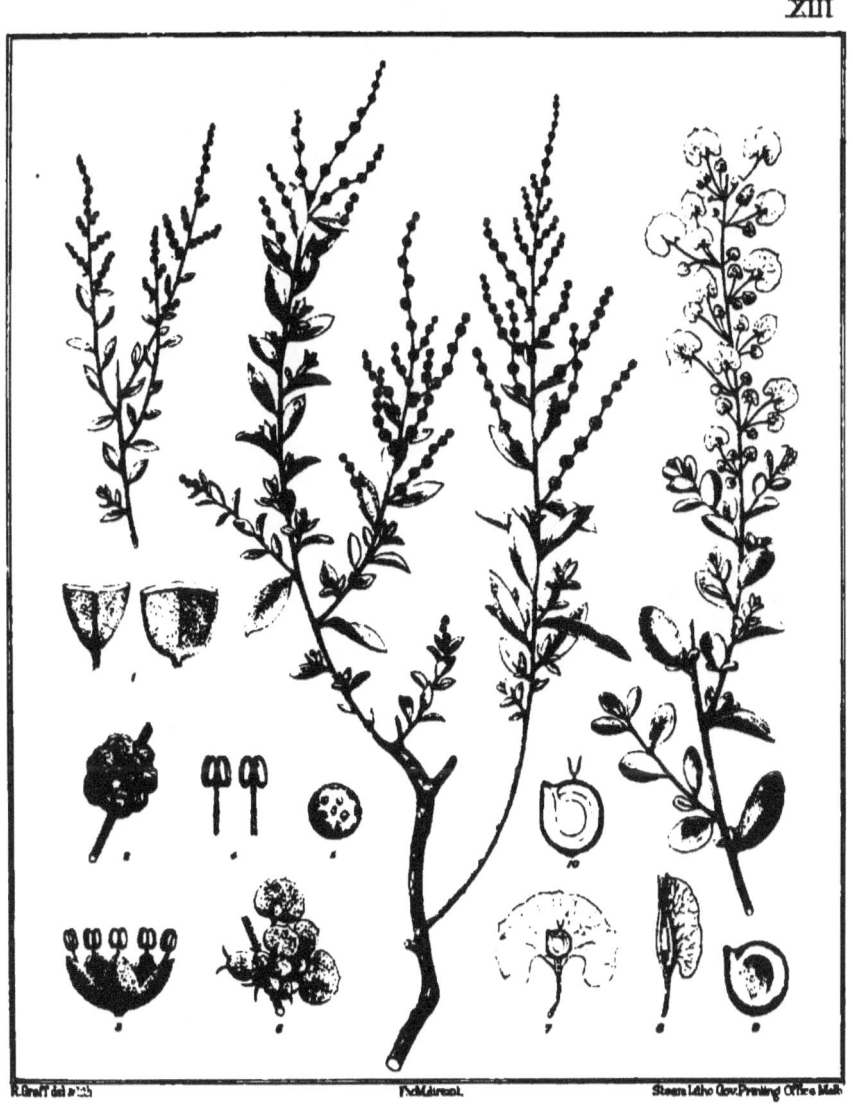

Atriplex stipitatum Bentham.

ATRIPLEX PALUDOSUM.

R. Brown, prodromus florae Novae Hollandiae 406 (1810).

PLATE XIV.

1, portion of two leaves.
2, cluster of unexpanded staminate flowers.
3, an unexpanded staminate flower, separated.
4, an expanded staminate flower.
5, front- and back-view of a stamen.
6, pollen-grain.
7, two pistillate flowers.
8, five fruit-bearing calyces.
9, a fruit with half its calyx removed.
10, a seed.
11, transverse section of a seed.
12, longitudinal section of a fruit.

All enlarged, but to various extent.

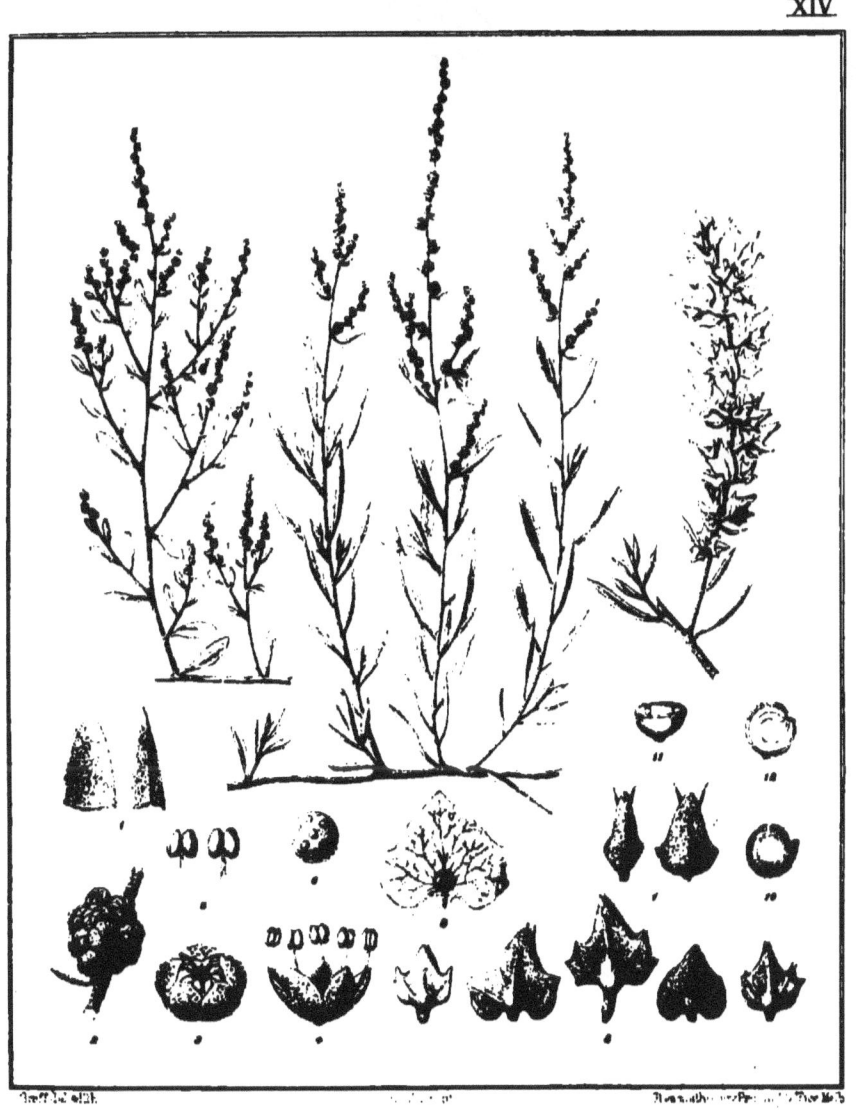

Atriplex paludosum R.Brown

ATRIPLEX CINEREUM.

Poiret, Encyclopédie méthodique, suppl. i. 471 (1810).

PLATE XV.

1, portion of a leaf.

2 and 3, staminate flowers.

4, front- and back-view of a stamen.

5, pollen-grain.

6, four pistillate flowers.

7, a pistillate flower, one half of the calyx removed.

8, three fruit-bearing calyces.

9, a fruit-bearing calyx, one half of it removed.

10, a seed.

11, transverse section of a seed.

12, longitudinal section of a seed.

All enlarged, but to various extent.

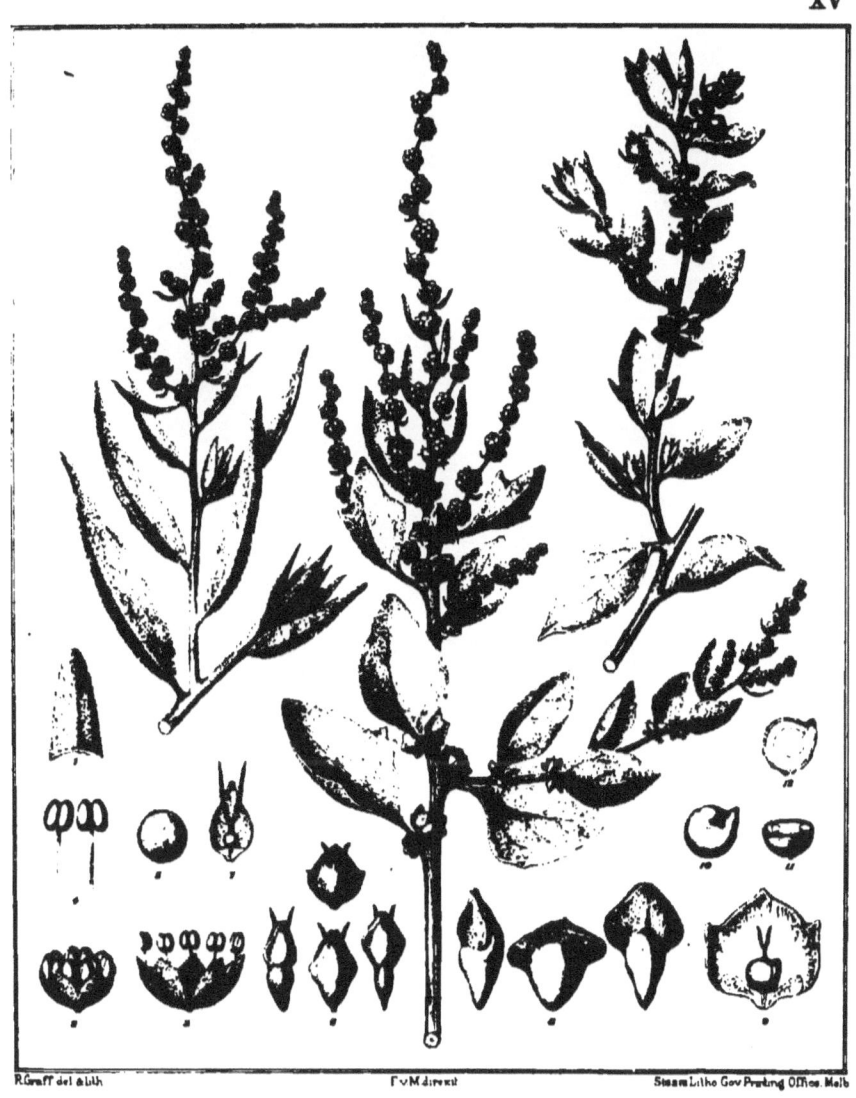

Atriplex cinereum *Poiret*

ATRIPLEX NUMMULARIUM.

Lindley in Mitchell's Tropical Australia 64 (1848).

PLATE XVI.

1, portion of a branchlet with unexpanded staminate flowers.
2, an expanded staminate flower.
3, front- and back-view of a stamen.
4, pollen-grain.
5, a pistillate flower.
6 and 7, fruit-bearing calyces.
8, a fruit-bearing calyx, one half of it removed.
9, longitudinal section of two fruit-bearing calyces.
10, a fruit, separated.
11, a seed.
12, transverse section of a seed.
13, longitudinal section of a fruit.

All enlarged, but to various extent.

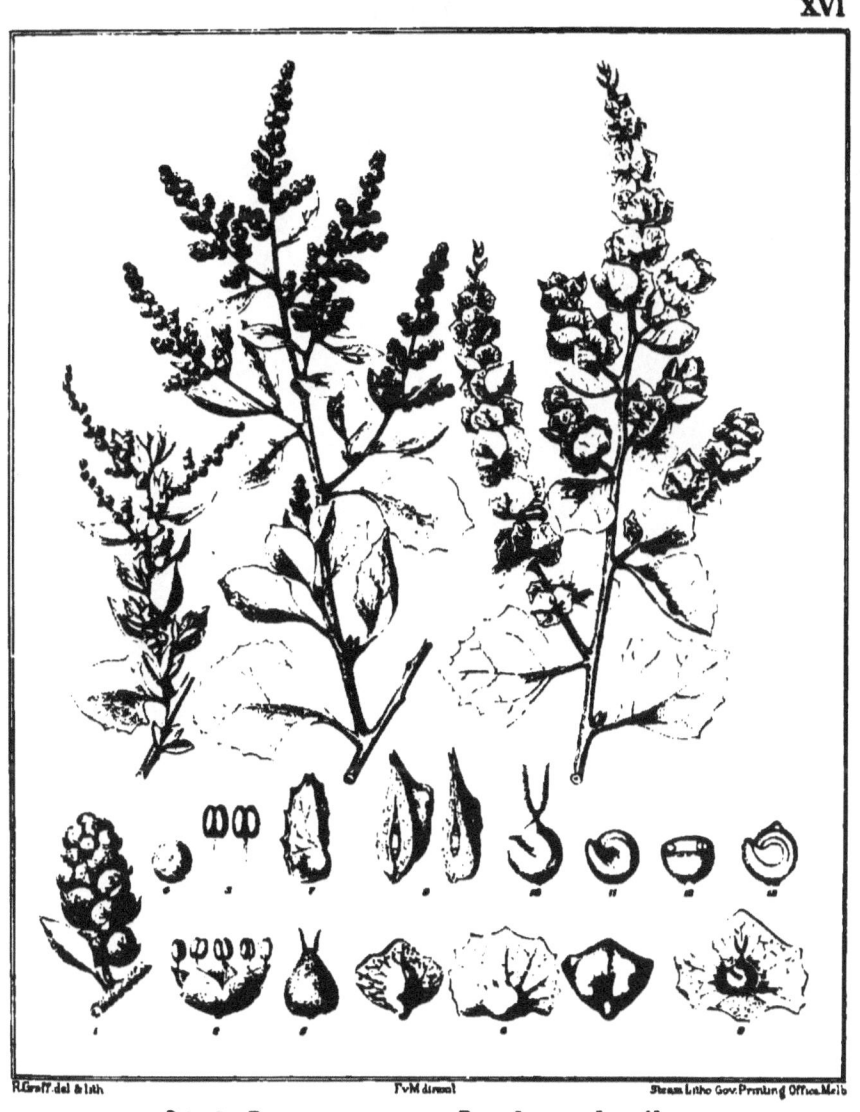

Atriplex nummularium Lindley.

ATRIPLEX HYMENOTHECUM.

Moquin in De Candolle prodromus xiii, pars ii, 101 (1849).

PLATE XVII.

1, portion of two leaves.

2, cluster of unexpanded staminate flowers.

3, an expanded staminate flower.

4, front- and back-view of a stamen.

5, pollen-grain.

6, a pistillate flower.

7, a pistillate flower, one half of the calyx removed.

8, a fruit.

9, a seed.

10, transverse section of a seed.

11, longitudinal section of a seed.

All enlarged, but to various extent.

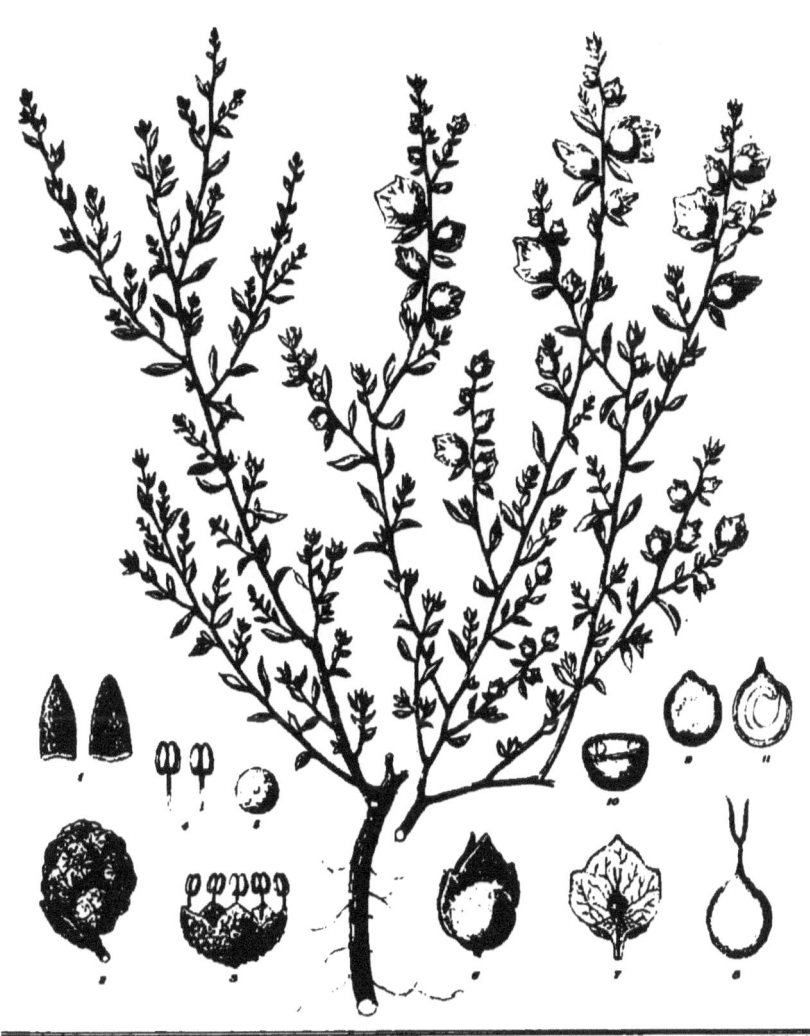

Atriplex hymenothecum *Moquin.*

ATRIPLEX VESICARIUM.

Heward in Bentham's Flora Australiensis v, 172 (1870).

PLATE XVIII.

1, portion of a branchlet with a leaf.
2, a staminate flower.
3, front- and back-view of a stamen.
4, pollen-grain.
5, pistillate flower, one half of the calyx removed.
6, three pistillate flowers.
7, two fruits, one half of the calyx removed.
8, longitudinal section of a fruit and its calyx.
9, a fruit, separated.
10, a seed.
11, longitudinal section of a fruit.

All enlarged, but to various extent.

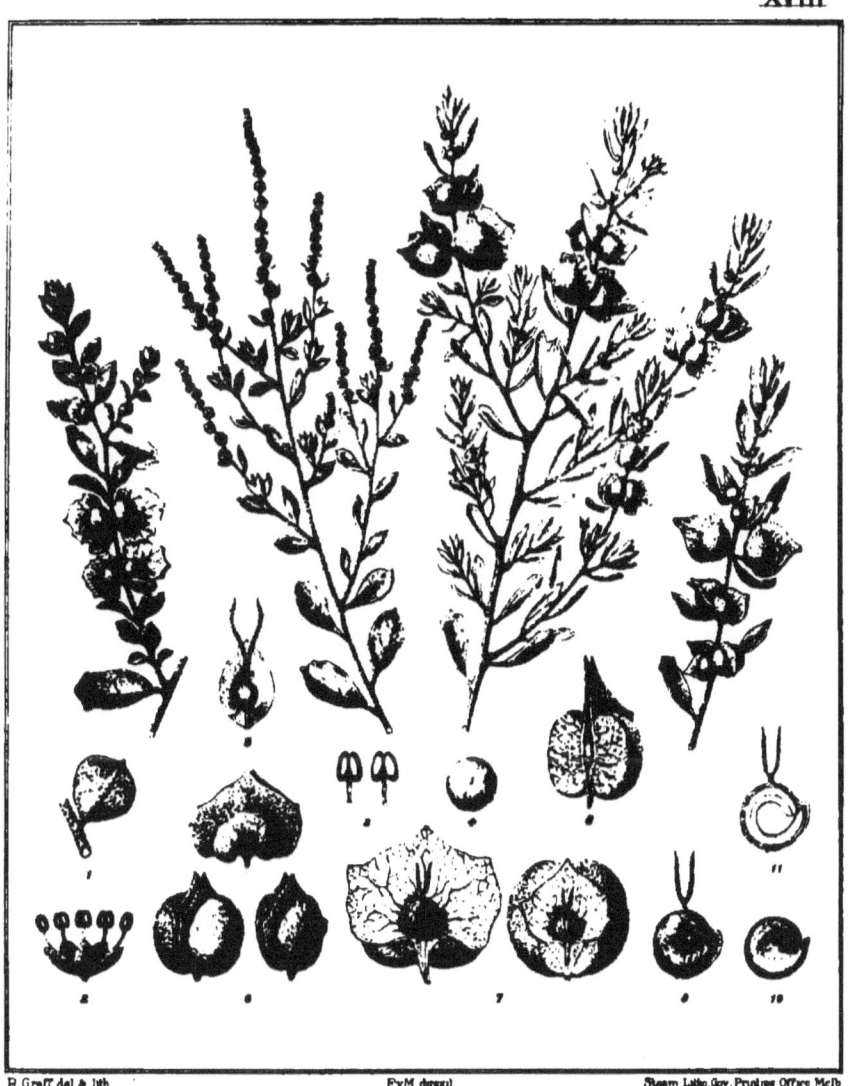

Atriplex vesicarium Heward

ATRIPLEX HALIMOIDES.

Lindley in Mitchell's Three Expeditions i, 282 (1838).

PLATE XIX.

1, staminate flower.

2, front- and back-view of a stamen.

3, pollen-grain.

4, three pistillate flowers.

5, fruit-bearing calyx, one half of it removed.

6, longitudinal section of a fruit-bearing calyx.

7, a fruit, separated.

8, a seed.

9, transverse section of a seed.

10, longitudinal section of a seed.

All enlarged, but to various extent.

XIX

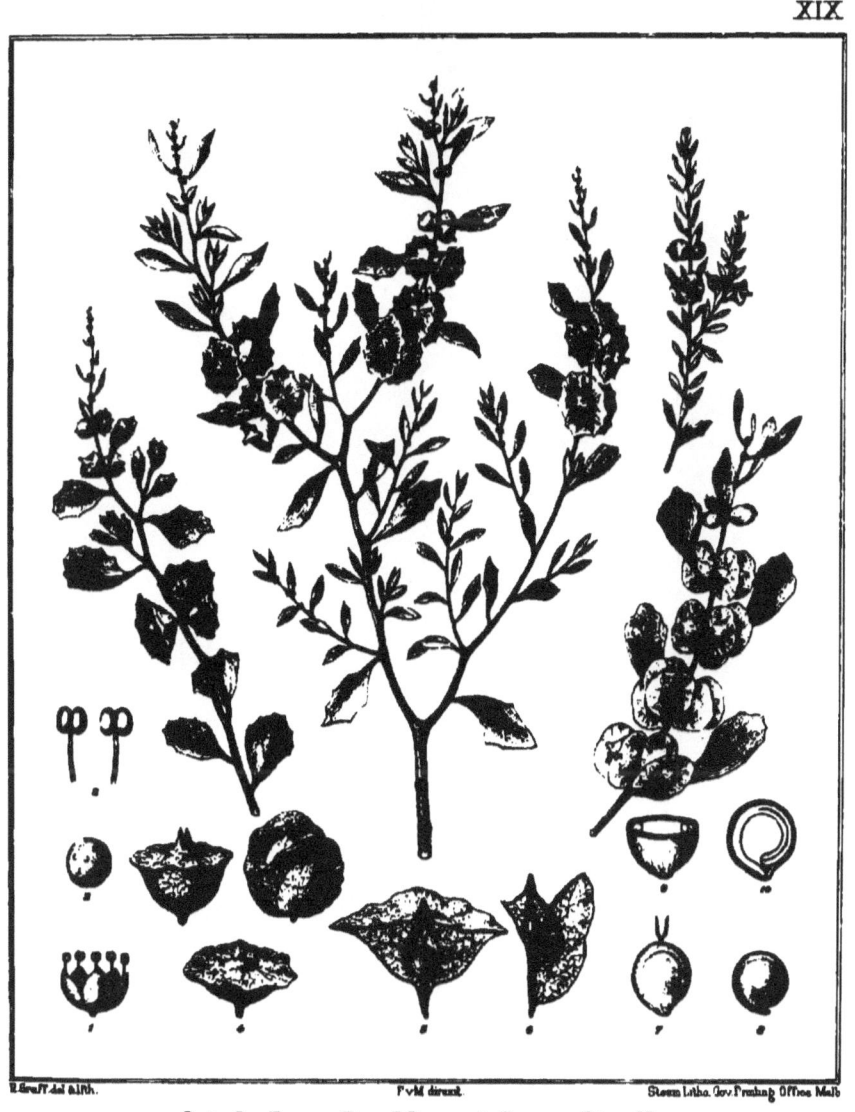

Atriplex halimoides *Lindley*

ATRIPLEX SPONGIOSUM.

F. v. Mueller in Transactions of the Philosophical Institute of Victoria ii, 74 (1857).

PLATE XX.

1, cluster of staminate and pistillate flowers.

2, two expanded staminate flowers.

3, back- and front-view of a stamen.

4, pollen-grain.

5, two pistillate flowers.

6, four fruit-bearing calyces.

7, transverse section of a fruit with its calyx.

8, fruit with one half of the calyx removed.

9, a fruit, separated.

10, a seed.

11, transverse section of a fruit.

12, longitudinal section of a fruit.

All enlarged, but to various extent.

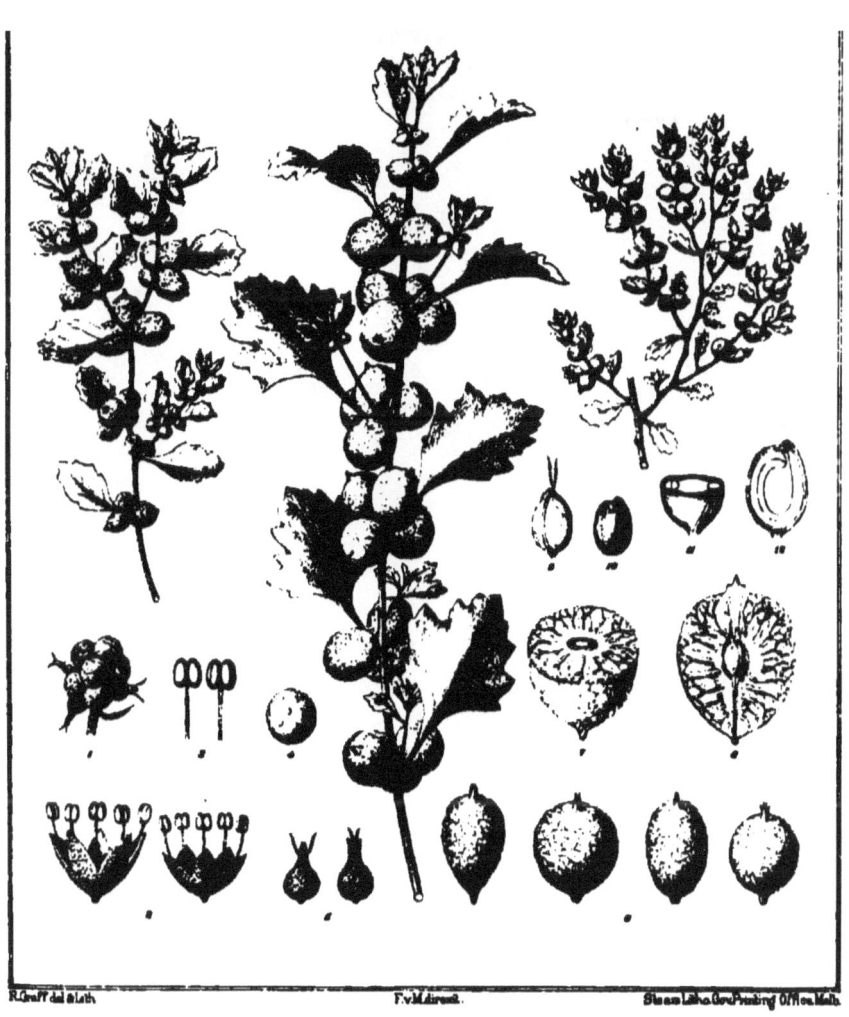

Atriplex spongiosum *F.vM.*

WILLIAM A. SETCHELL,
UNIV. OF CALIFORNIA,
BERKELEY, . . . CALIF.

ICONOGRAPHY

OF

AUSTRALIAN SALSOLACEOUS PLANTS.

BY

BARON FERD. VON MUELLER, K.C.M.G., M. & PH.D., F.R.S.,

GOVERNMENT BOTANIST OF THE COLONY OF VICTORIA.

THIRD DECADE.

By Authority:
ROBERT S. BRAIN, GOVERNMENT PRINTER, MELBOURNE.
1890.

Rhagodia Billardieri.

R. Brown, prodromus florae Novae Hollandiae 408 (1810).

PLATE XXI

1, portion of a branchlet with flowers.

2, a staminate flower.

3, a flower with both stamens and pistil.

4, front- and back-view of a stamen.

5, pollen-grain.

6, a young fruit.

7, a mature fruit.

8, longitudinal section of a fruit.

9, a seed.

10, longitudinal section of a seed.

All enlarged, but to various extent.

XXI

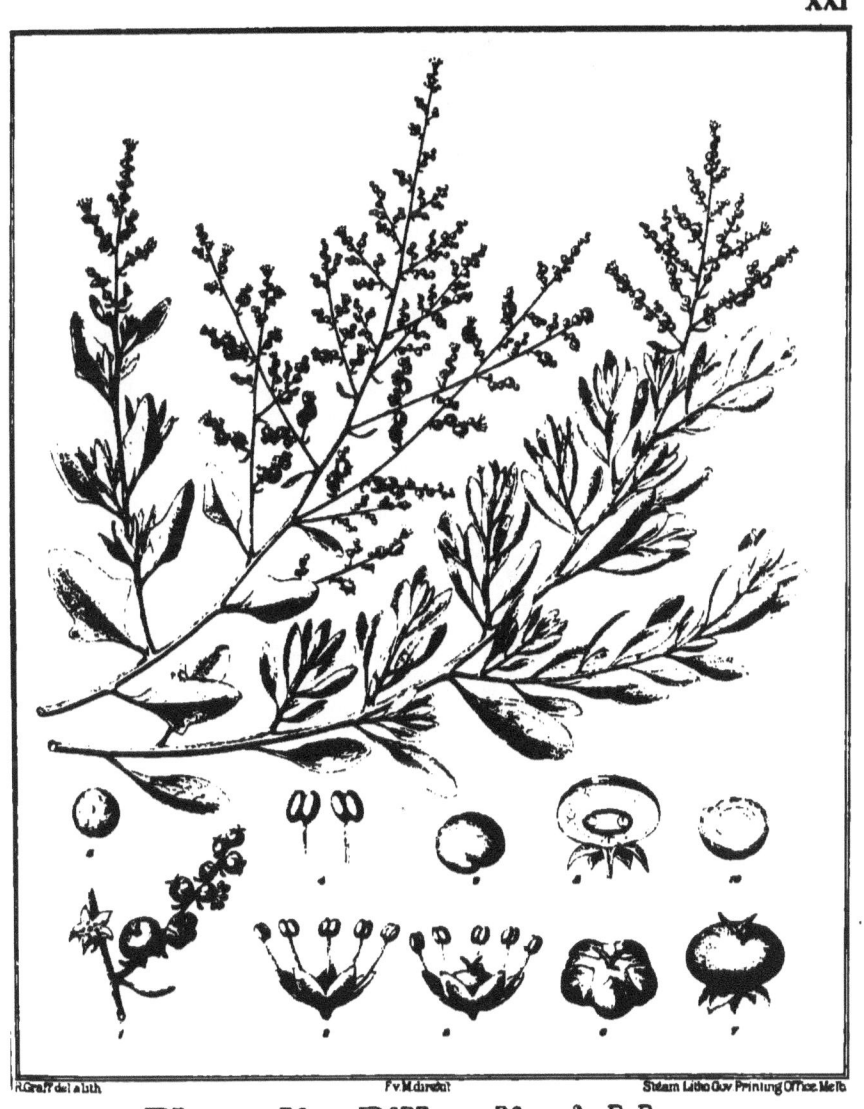

Rhagodia Billardieri *R.Brown.*

Rhagodia spinescens.

R. Brown, prodromus florae Novae Hollandiae 408 (1810).

PLATE XXII

1, portions of two branchlets with flowers.
2, three expanded flowers with various numbers of stamens.
3, front- and back-view of a stamen.
4, pollen-grain.
5, a young fruit.
6, a mature fruit.
7, longitudinal section of a fruit.
8, a seed.
9, longitudinal section of a seed.

All enlarged, but to various extent.

XXII

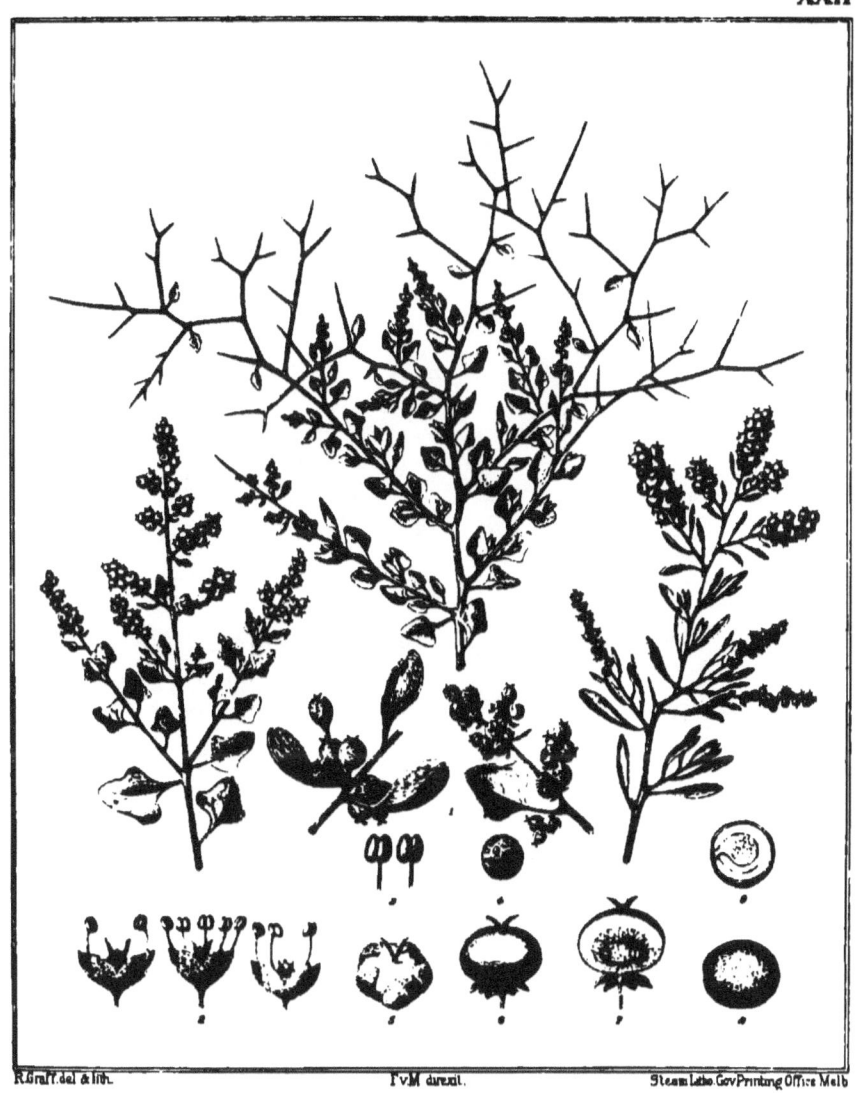

Rhagodia linifolia.

R. Brown, prodromus florae Novae Hollandiae 408 (1810).

PLATE XXIII.

1, portion of a branchlet with flowers.

2, a staminate flower.

3, front- and back-view of a stamen.

4, pollen-grain.

5, a flower with stamen and pistil, both perfect.

6, a young fruit.

7, a mature fruit.

8, longitudinal section of a fruit.

9, a seed.

10, transverse section of a seed.

11, longitudinal section of a seed.

All enlarged, but to various extent.

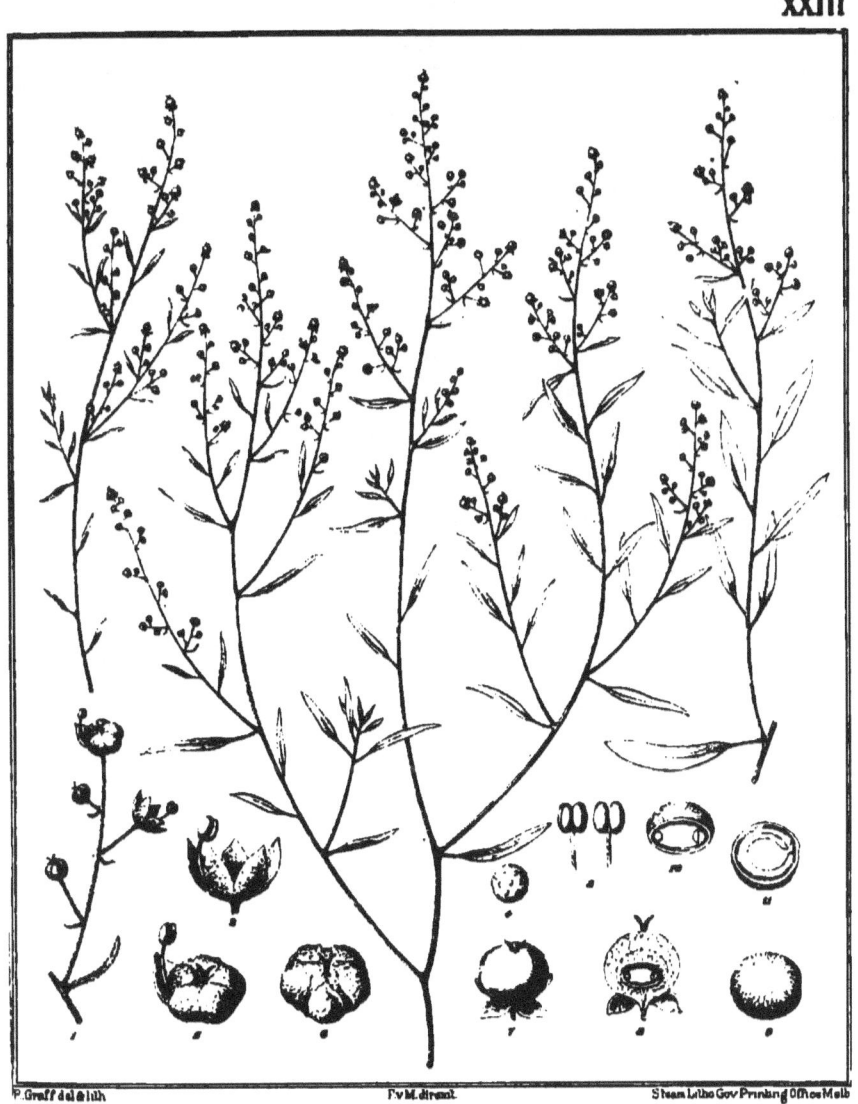

Rhagodia linifolia R.Brown.

Rhagodia nutans.

R. Brown, prodromus florae Novae Hollandiae 408 (1810).

PLATE XXIV.

1, portions of two branchlets with flowers.

2, two staminate flowers.

3, front- and back-view of a stamen.

4, pollen-grain.

5, a flower with stamen and pistil, both perfect.

6, a fruit.

7, longitudinal section of a fruit.

8, a seed.

9, transverse section of a seed.

10, longitudinal section of a seed.

All enlarged, but to various extent.

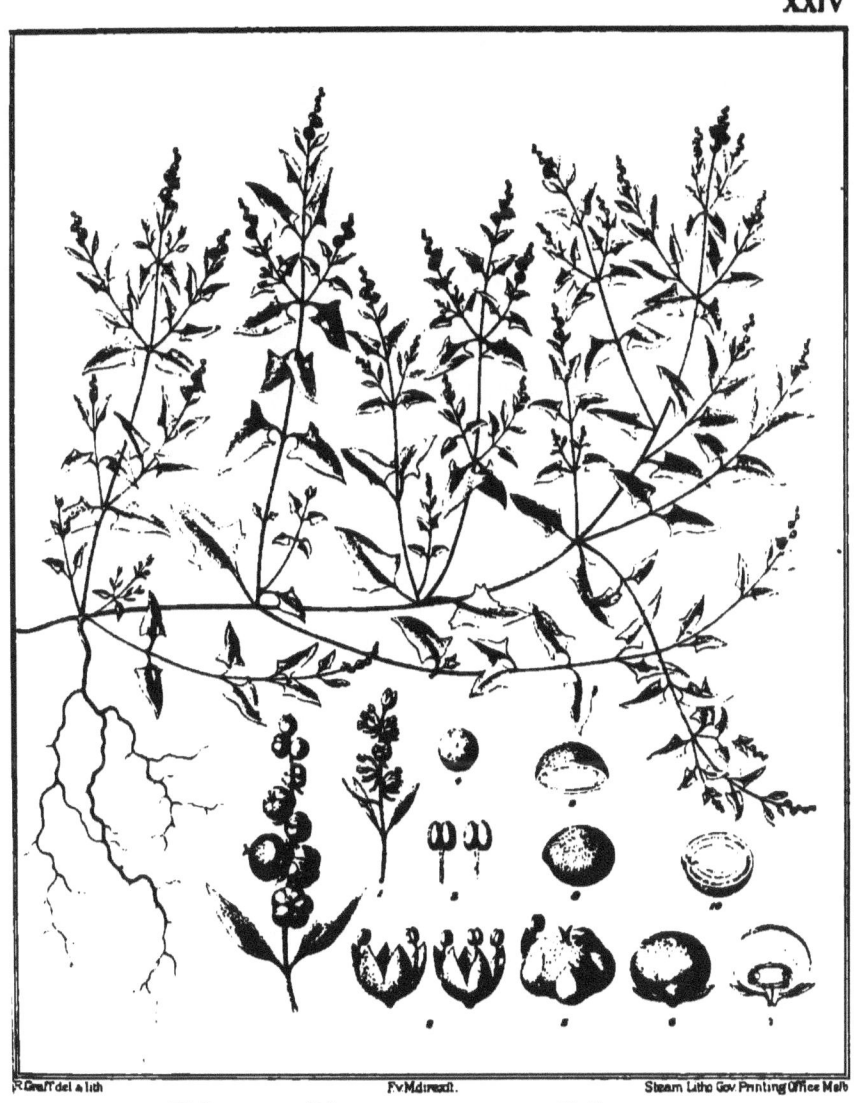

Rhagodia nutans *R. Brown.*

Rhagodia hastata.

R. Brown, prodromus florae Novae Hollandiae 408 (1810).

PLATE XXV.

1, portion of a branchlet with flowers.

2, a staminate flower.

3, front- and back-view of a stamen.

4, pollen-grain.

5, a flower with stamen and pistil perfect.

6, a fruit.

7, a seed.

8, transverse section of a fruit.

9, longitudinal section of a seed.

All enlarged, but to various extent.

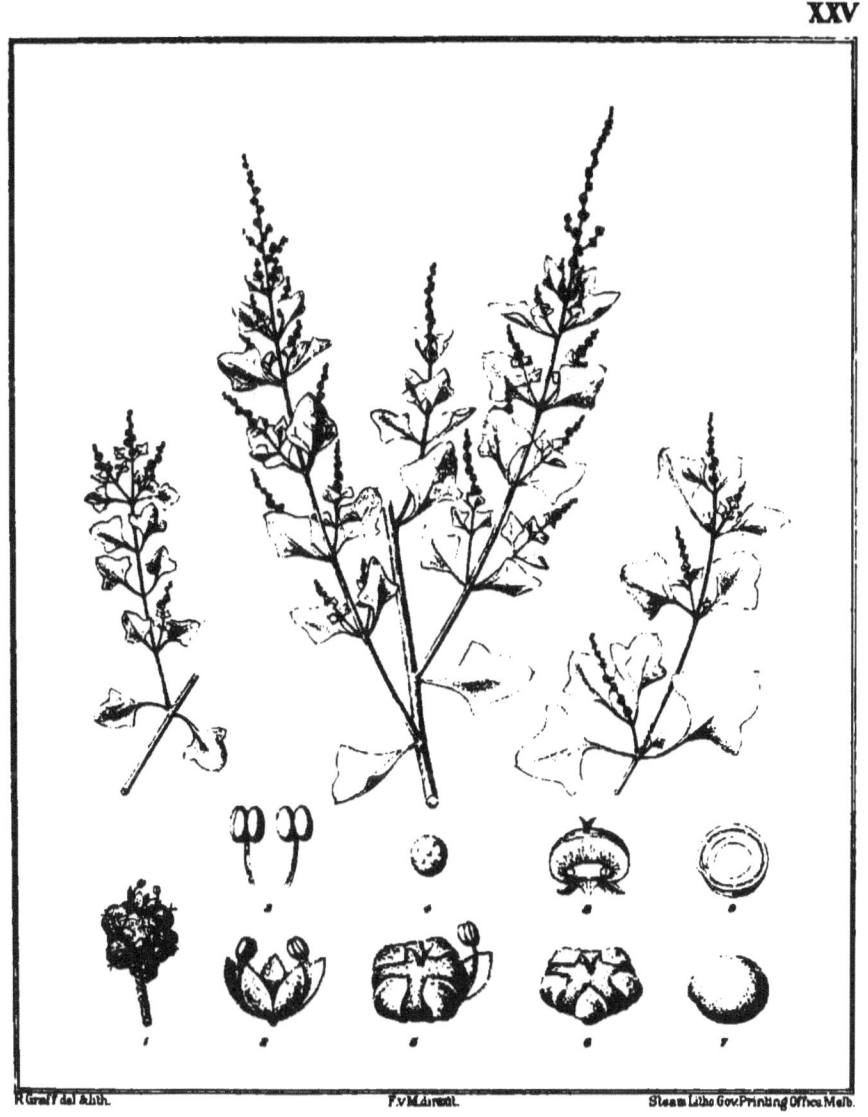

Rhagodia hastata *R.Brown.*

Chenopodium triangulare.

R. Brown, prodromus florae Novae Hollandiae 407 (1810).

PLATE XXVI.

1, portion of a branchlet with flowers.

2, a flower, part of the calyx removed.

3, a flower, stamen and pistil perfect.

4, front- and back-view of a stamen.

5, pollen-grain.

6, a young fruit.

7, a mature fruit.

8, a fruit, seen from beneath.

9, longitudinal section of a fruit.

10, a seed.

11, longitudinal section of a seed.

All enlarged, but to various extent.

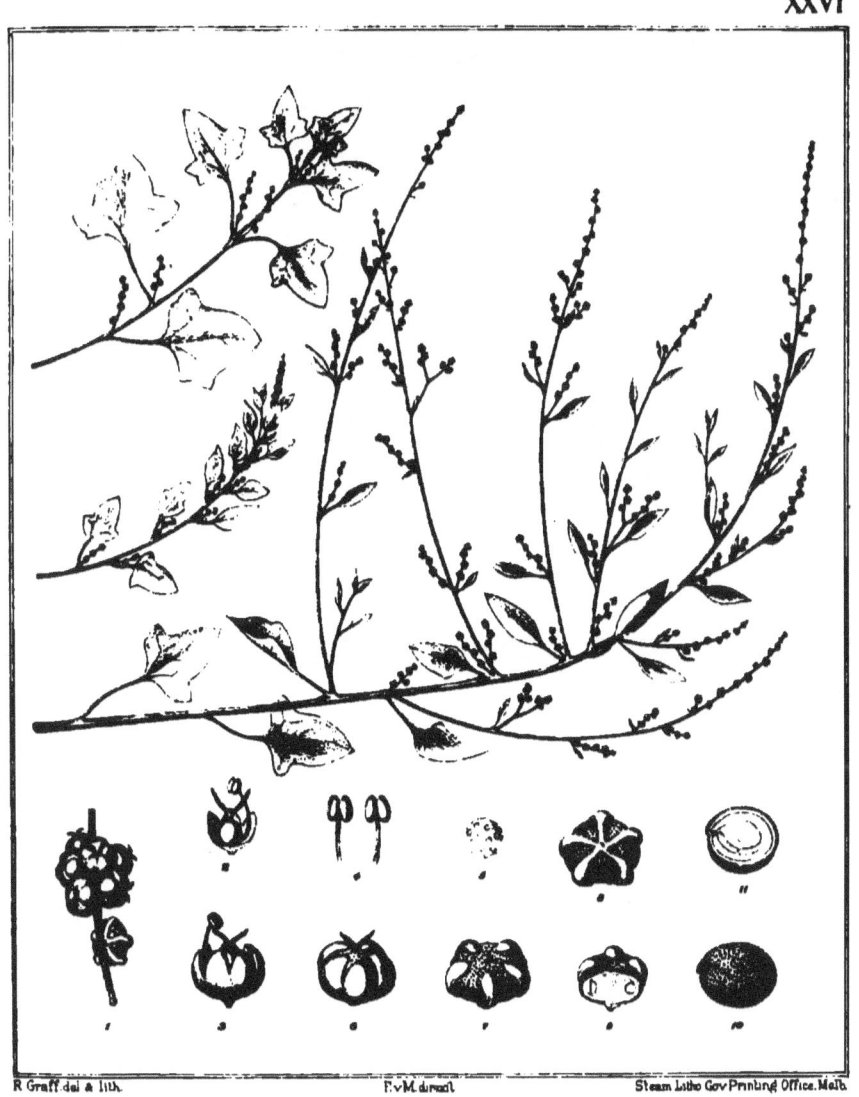

Chenopodium triangulare R. Brown.

CHENOPODIUM MICROPHYLLUM.

F. v. M. in the Transactions of the Philosophical Institute of Victoria ii, 74 (1857).

PLATE XXVII.

1, three different leaves.

2, portion of a branchlet with flowers.

3, a young flower.

4, an expanded flower.

5, front- and back-view of a stamen.

6, pollen-grain.

7 and 8, two fruits.

9, a fruit, the calyx removed.

10, a seed.

11, transverse section of a fruit.

12, longitudinal section of a fruit.

All enlarged, but to various extent.

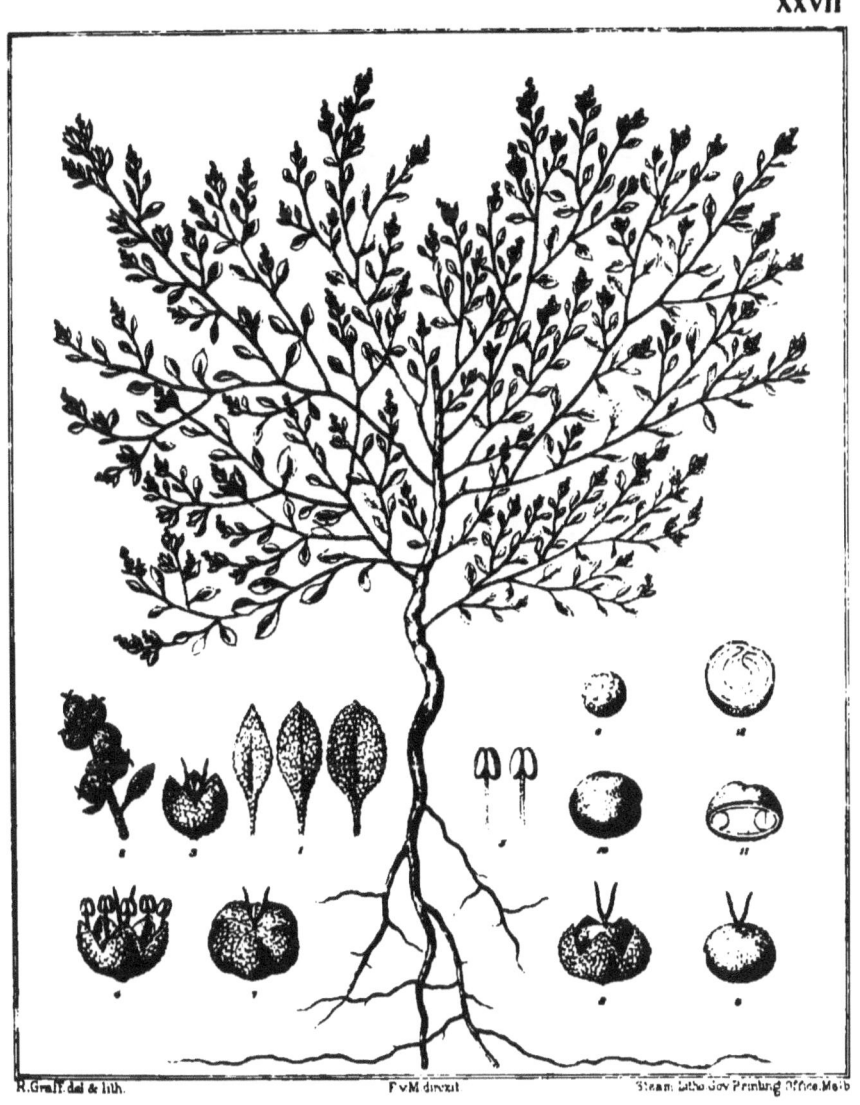

Chenopodium microphyllum F.v.M.

CHENOPODIUM NITRARIACEUM.

F. v. M. in Bentham's Flora Australiensis v, 158 (1870).

PLATE XXVIII.

1, portion of a branchlet with flowers.

2, two young flowers.

3, expanded flower, part of the calyx removed.

4, expanded flower.

5, front- and back-view of a stamen.

6, pollen-grain.

7, a fruit.

8, longitudinal section of a fruit.

9, a fruit, the calyx removed.

10, a seed.

11, longitudinal section of a fruit.

All enlarged, but to various extent.

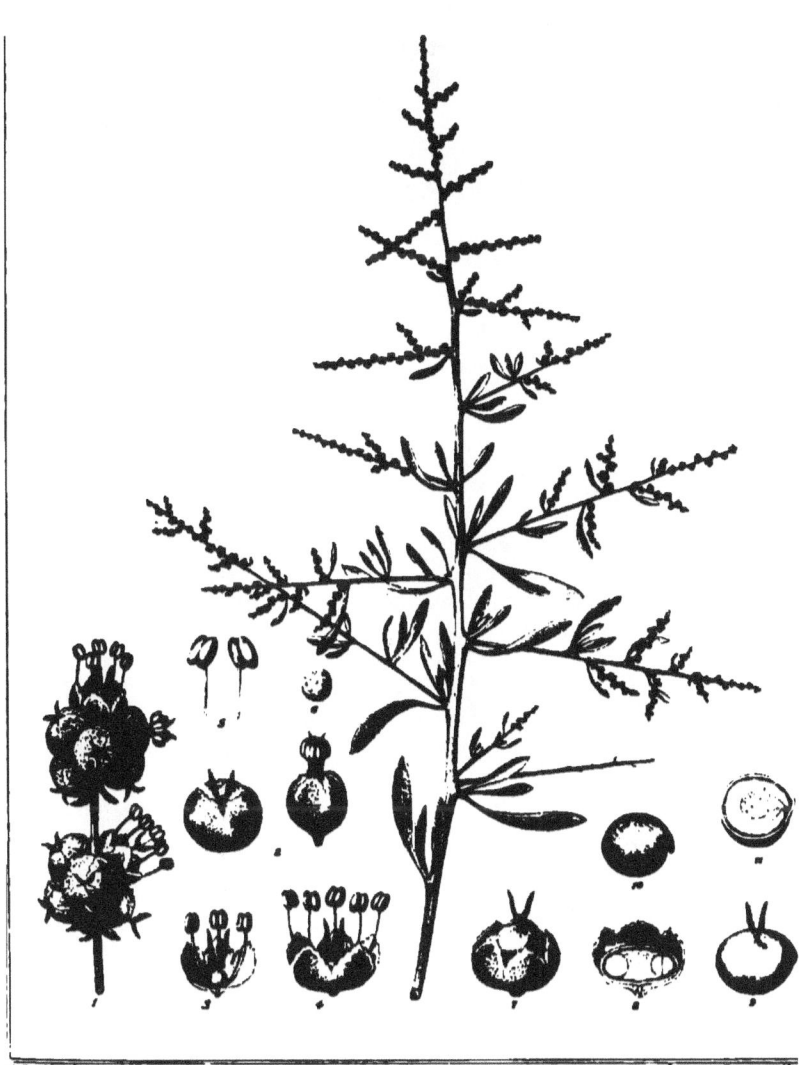

Chenopodium nitrariaceum F.v.M

CHENOPODIUM AURICOMUM.

Lindley in Mitchell's Tropical Australia 94 (1848).

PLATE XXIX.

1, a young flower.
2 and 3, expanded flowers.
4, front- and back-view of a stamen.
5, pollen-grain.
6, two fruits at different stages of age.
7, a seed.
8, transverse section of a seed.
9, longitudinal section of a seed.

All enlarged, but to various extent.

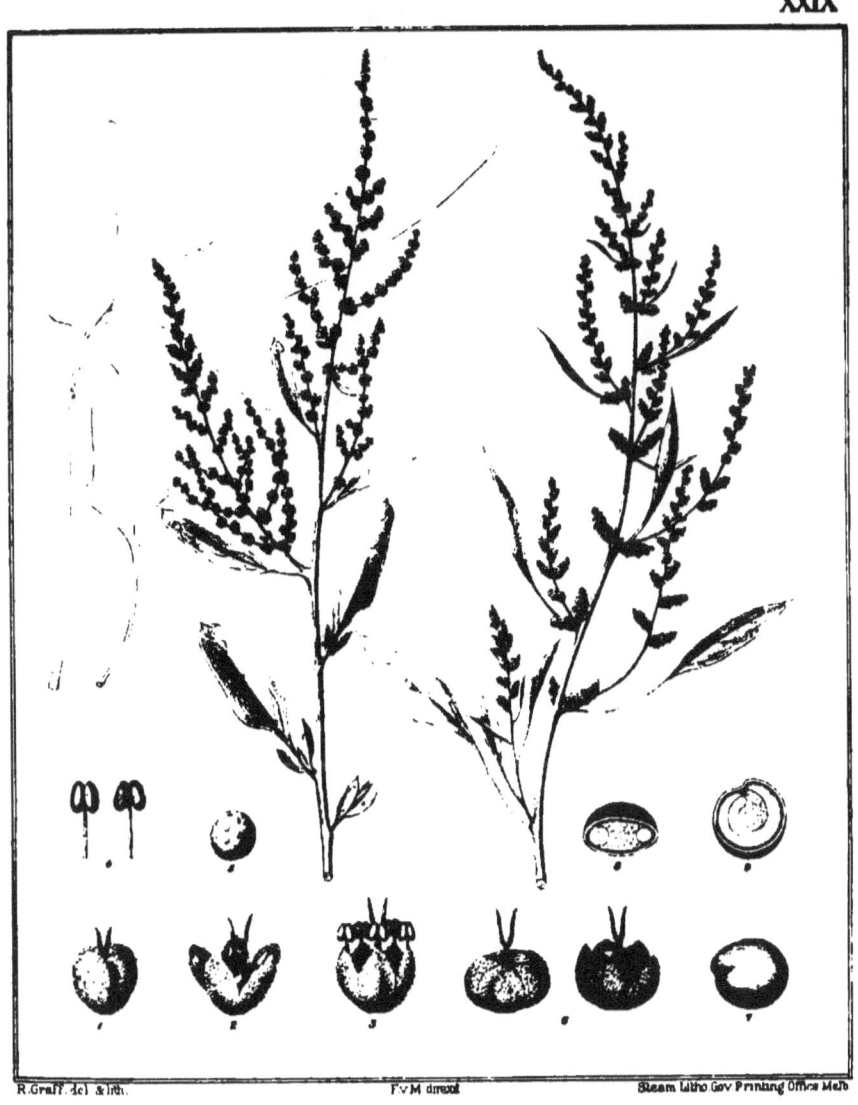

Chenopodium auricomum Lindley

CHENOPODIUM ATRIPLICINUM.

F. v. M. fragmenta phytographiae Australiae vii, 11 (1869).

PLATE XXX.

1, portion of a branchlet with flowers.
2, a flower, part of the calyx removed.
3, front- and back-view of a stamen.
4, pollen-grain.
5 and 6, three fruits.
7, longitudinal section of a fruit.
8, a fruit, the calyx removed.
9, transverse section of a fruit.
10, longitudinal section of a fruit.

All enlarged, but to various extent.

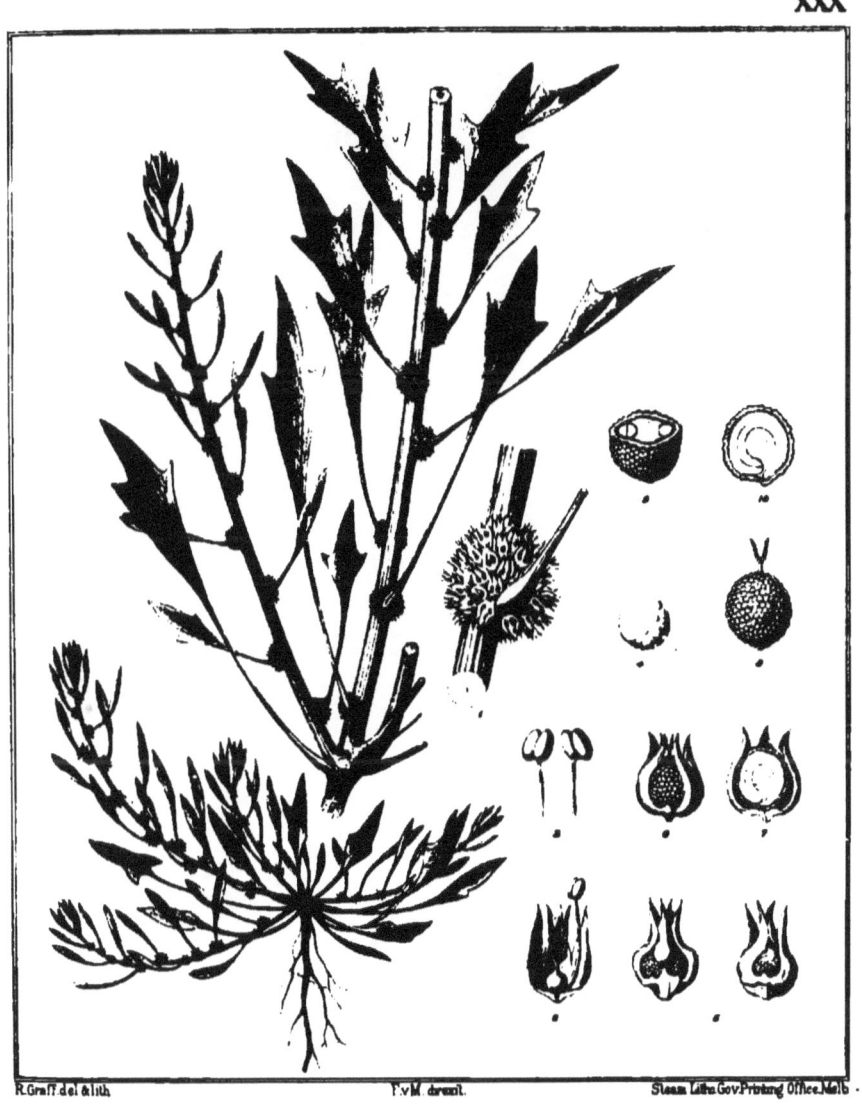

Chenopodium atriplicinum F.v.M.

WILLIAM A. SETCHELL,
UNIV. OF CALIFORNIA,
BERKELEY, - - - CALIF.

ICONOGRAPHY

OF

AUSTRALIAN SALSOLACEOUS PLANTS.

BY

BARON FERD. VON MUELLER, K.C.M.G., M. & PH.D., F.R.S.,

GOVERNMENT BOTANIST OF THE COLONY OF VICTORIA.

"Mulin unde suum vocant, passo vero gelastrl."—*Plinus, Natlh. xxii., 14.*

FOURTH DECADE.

By Authority:
ROBERT S. BRAIN, GOVERNMENT PRINTER, MELBOURNE.
1890.

CHENOPODIUM CRISTATUM.

F. v. M. fragmenta phytographiae Australiae vii, 11 (1869)

PLATE XXXI.

1 and 2, portion of a branchlet with flowers and fruits

3, two separate flowers.

4, front- and back-view of a stamen.

5, pollen-grain.

6, three fruits with calyx.

7, a fruit, part of the calyx removed.

8, a fruit, the calyx wholly removed.

9, a seed.

10, transverse section of a fruit.

11, longitudinal section of a fruit.

All enlarged, but to various extent.

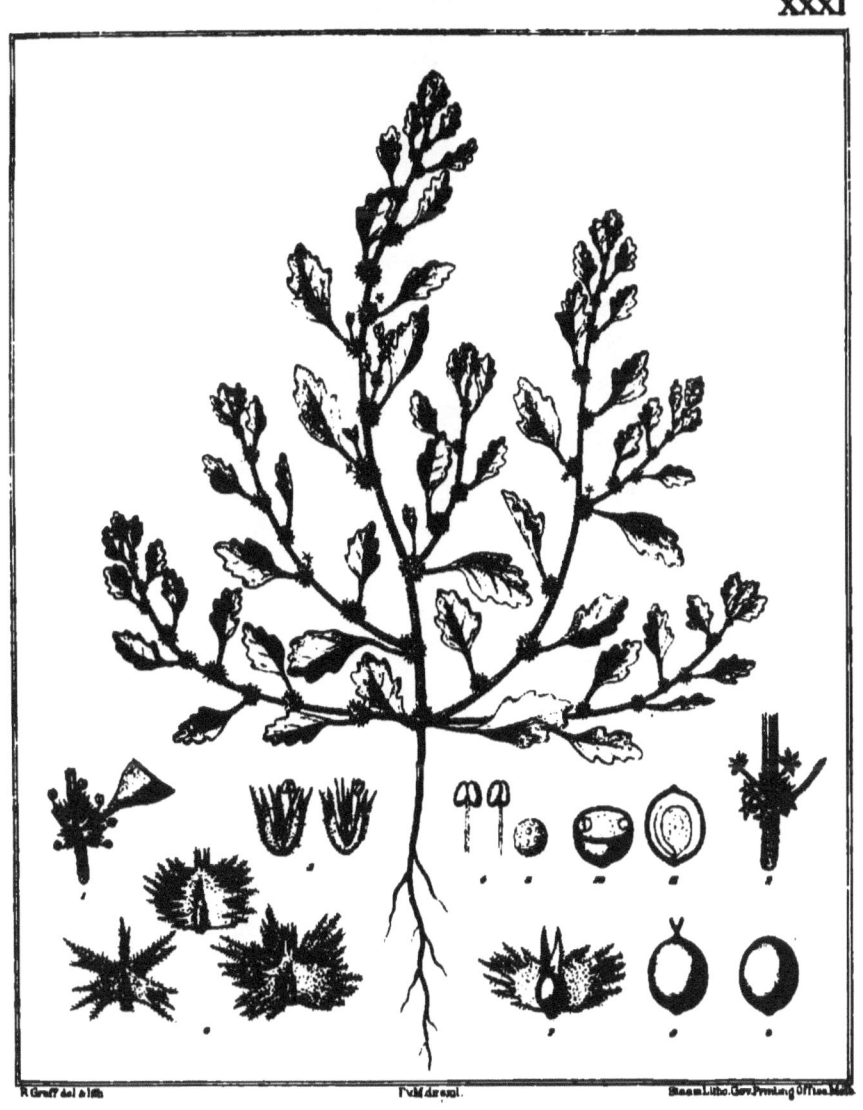

Chenopodium cristatum F.vM.

Chenopodium carinatum.

R. Brown, prodromus florae Novae Hollandiae 407 (1810).

PLATE XXXII.

1, portion of a branchlet with flowers.
2, a separate flower.
3, a flower, part of the calyx removed.
4, front- and back-view of a stamen.
5, pollen-grain.
6, 7 and 8, five fruits with their calyx.
9, a fruit, the calyx removed.
10, two seeds.
11, transverse section of two fruits.
14, longitudinal section of a fruit.

All enlarged, but to various extent.

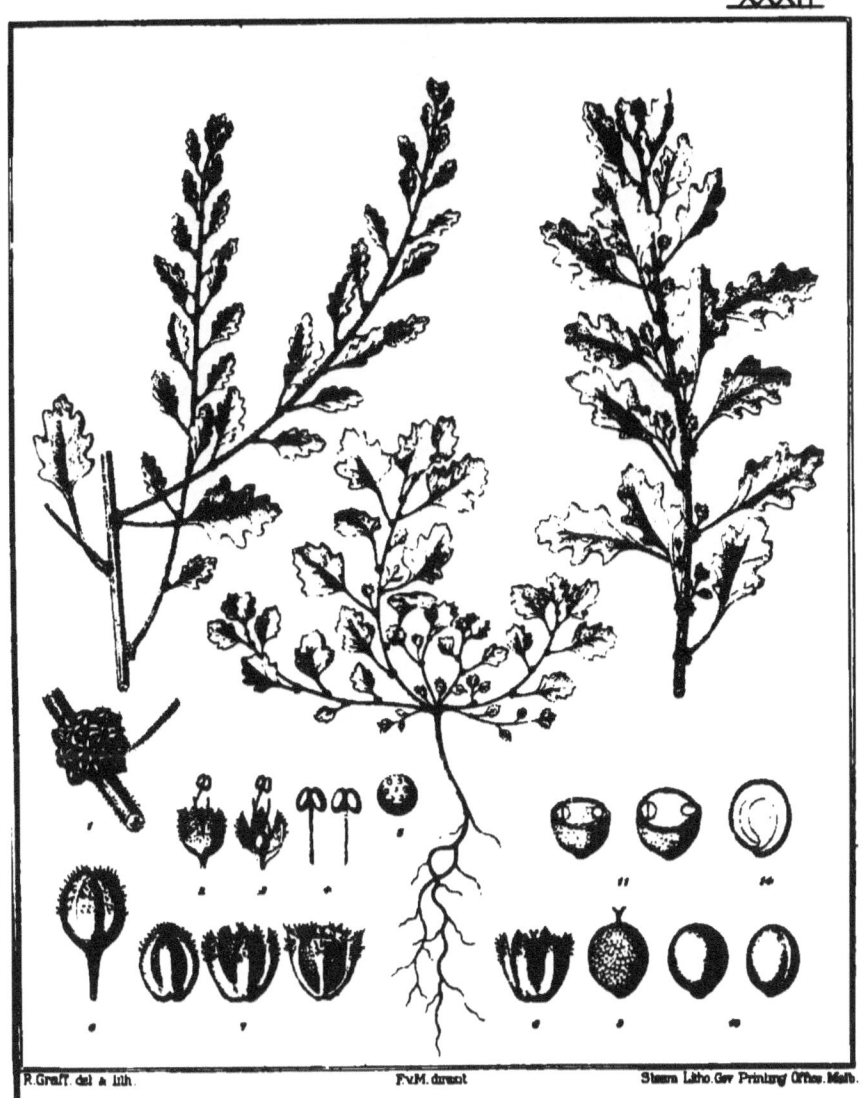

Chenopodium carinatum RBrown.

CHENOPODIUM RHADINOSTACHYUM.

F. v. M. in Wing's Southern Science Record (1882).

PLATE XXXIII.

1, portion of a branchlet with flowers.
2, an unexpanded flower.
3, 4 and 5, expanded flowers.
6, a flower, part of the calyx removed.
7, back- and front-view of a stamen.
8, pollen-grain.
9, a fruit.
10, a seed.
11, transverse section of a seed.
12, longitudinal section of a seed.

All enlarged, but to various extent.

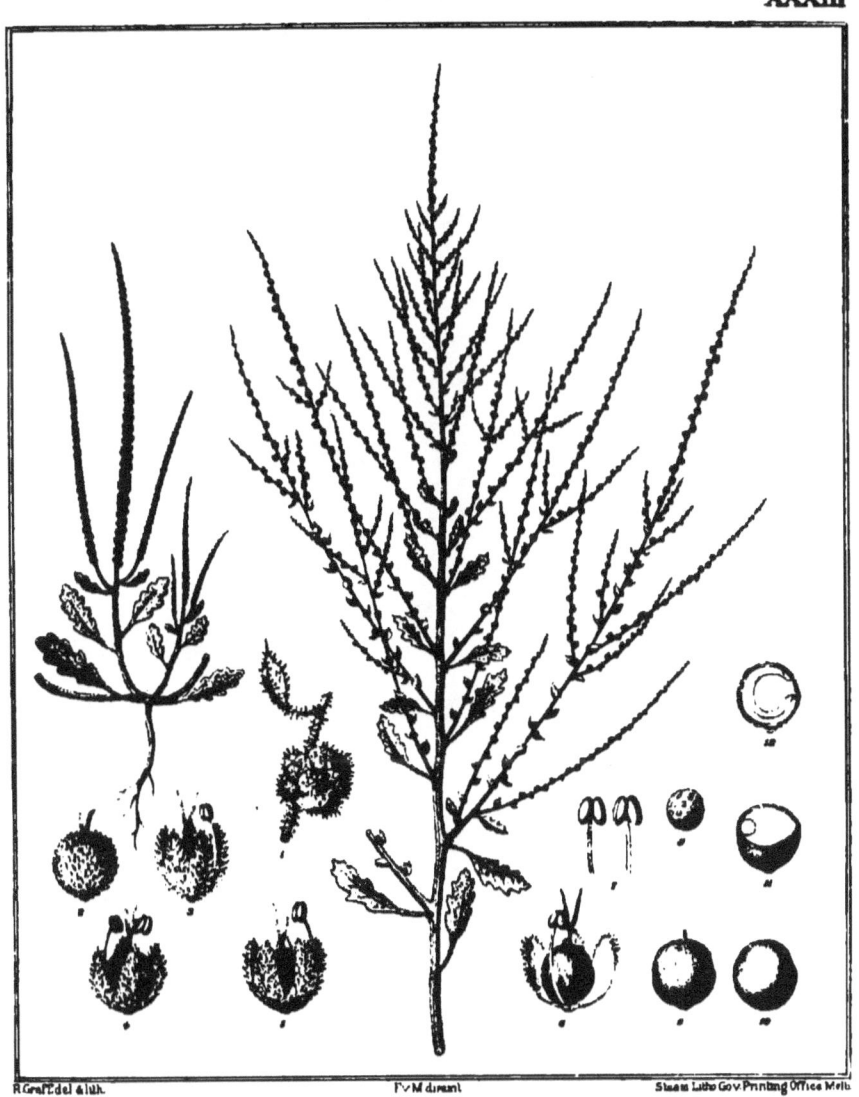

Chenopodium rhadinostachyum F.v.M.

DYSPHANIA SIMULANS.

F. v. M. and Tate in the Proceedings of the Royal Society of South Australia viii, 71 (1885).

PLATE XXXIV.

1, portion of a branchlet with unexpanded flowers.

2, portion of a branchlet with expanded flowers.

3, a detached flower.

4, a flower, part of the calyx removed.

5, front- and back-view of a stamen.

6, pollen-grain.

7 and 8, two fruits with their calyx.

9, a fruit, part of the calyx removed.

10, longitudinal section of a fruit with its calyx.

11, two fruits, the calyx removed, the style and stigmas remaining.

12, two seeds.

13, transverse section of a fruit.

14, longitudinal section of a fruit.

All enlarged, but to various extent.

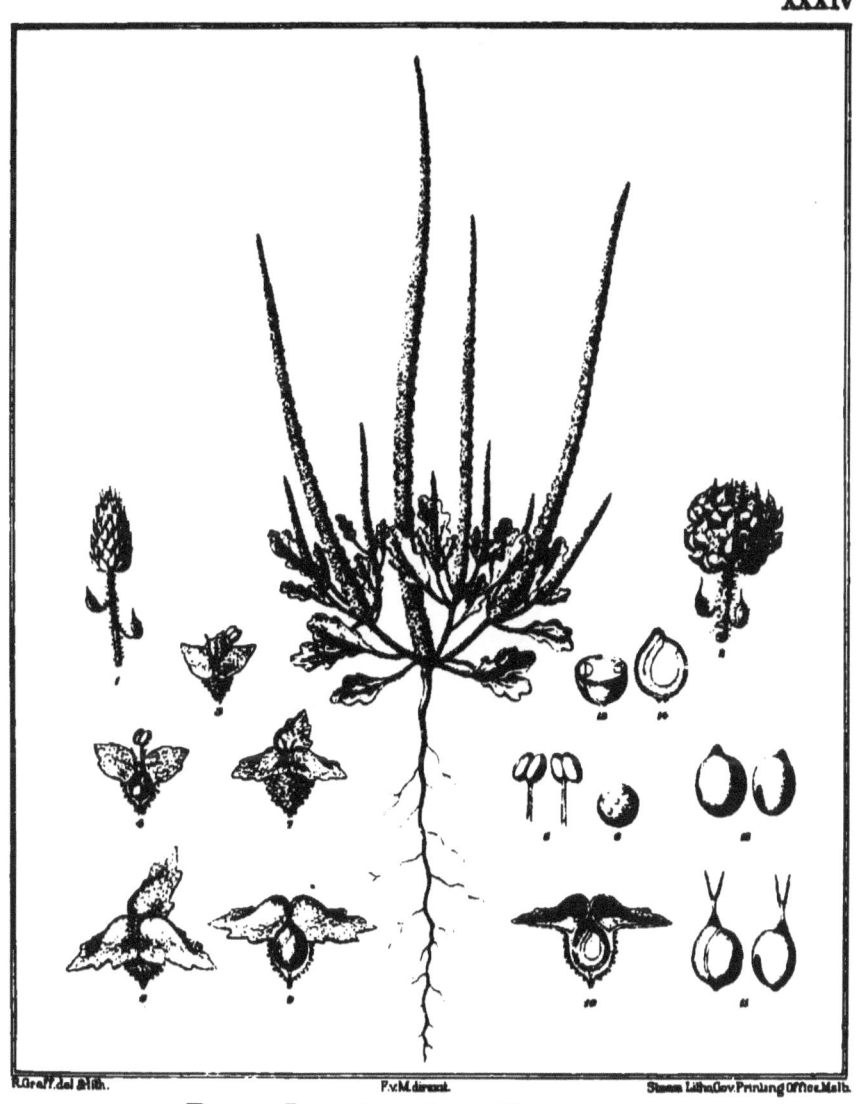

Dysphania simulans *FvM & Tate.*

Dysphania Plantaginella.

F. v. M. fragmenta phytographiae Australiae i, 61 (1858).

PLATE XXXV.

1, portion of a branchlet with bracts and unexpanded flowers.
2, portion of a branchlet with expanded flowers.
3, a detached flower.
4, a flower, part of the calyx removed.
5, back- and front-view of a stamen.
6, pollen-grain.
7, two fruits with their calyx.
8, transverse section of a fruit and calyx.
9, fruit with a two-lobed calyx.
10, two fruits, the calyx removed.
11, transverse section of a fruit.
12, longitudinal section of a fruit.

All enlarged, but to various extent.

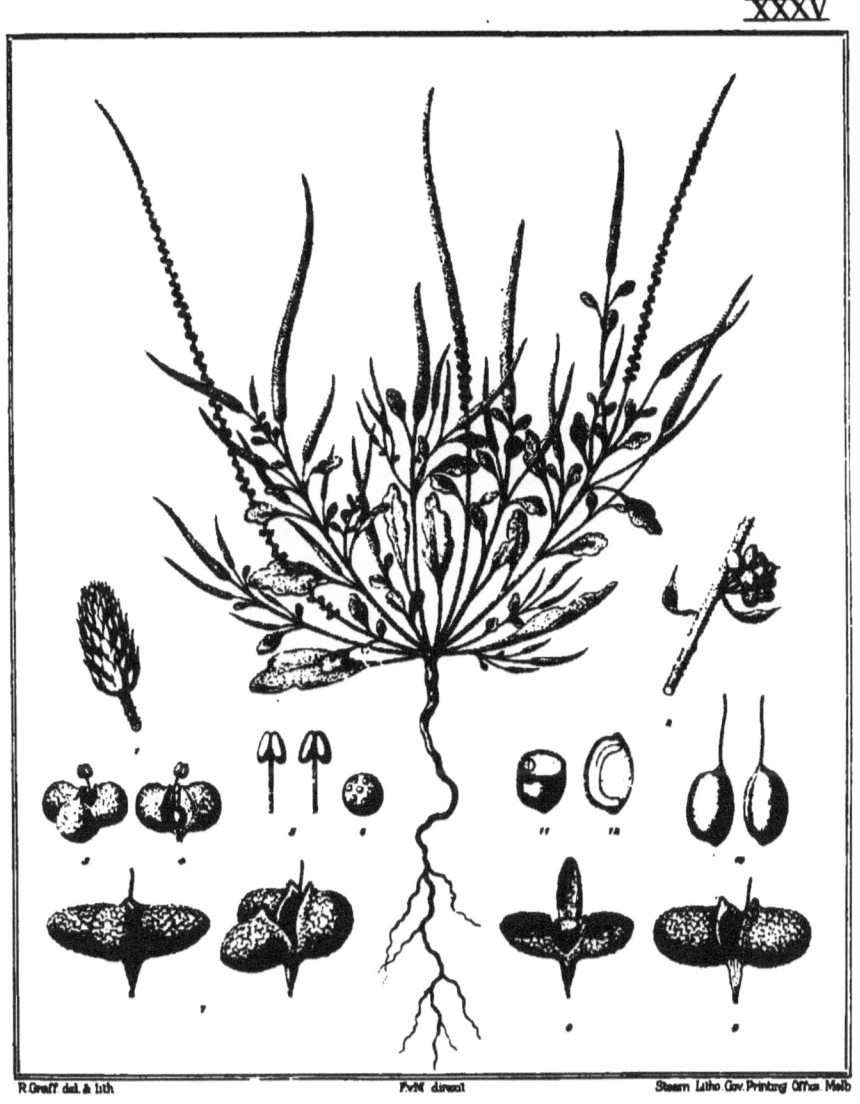

Dysphania Plantaginella *FvM*

Dysphania litoralis.

R. Brown, prodromus florae Novae Hollandiae 411 (1810).

PLATE XXXVI.

1, portion of a branchlet with flowers and a leaf.

2, two flowers.

3, a flower, part of the calyx removed.

4, front- and back-view of a stamen.

5, pollen-grain.

6, two young fruits.

7, transverse section of two fruits.

8, two mature fruits with their calyx.

9, a fruit, the calyx removed, the stigmas remaining.

10, a seed.

11, longitudinal section of a fruit.

12, transverse section of a fruit.

All enlarged, but to various extent.

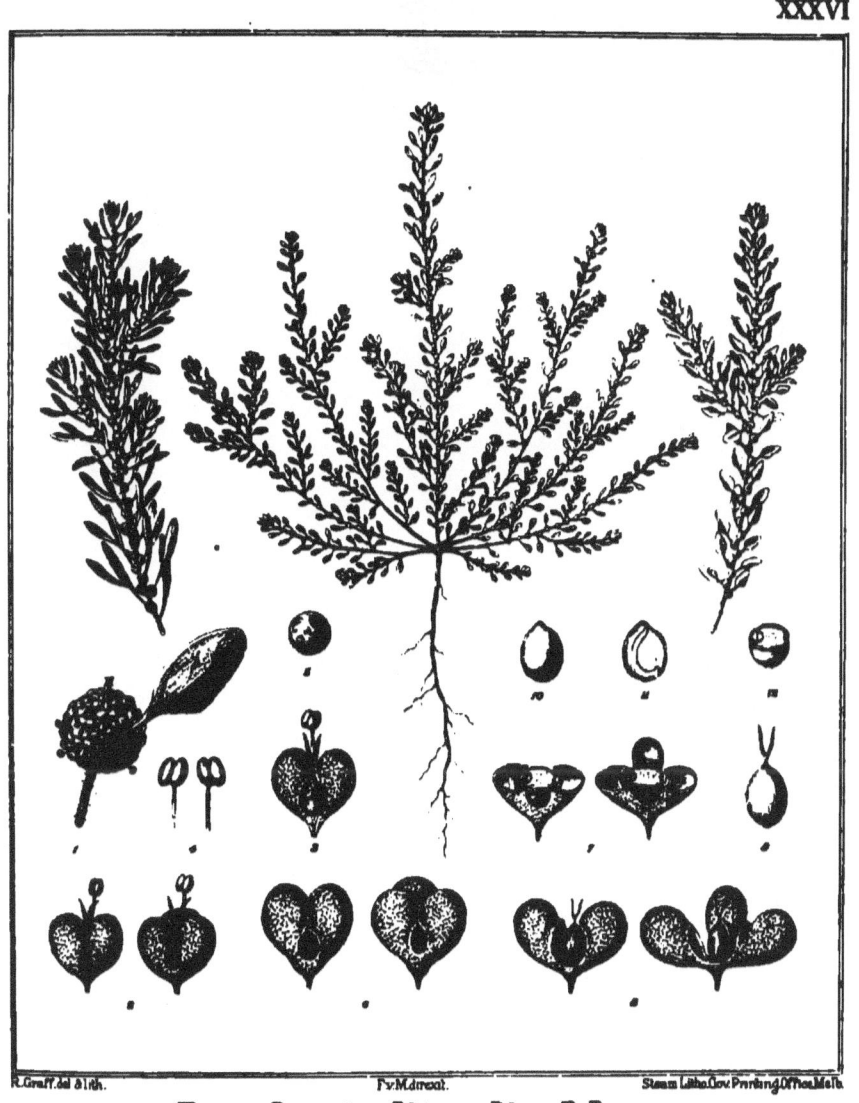

Dysphania litoralis R.Brown.

BABBAGIA DIPTEROCARPA.

F. v. M., Report on Plants of Babbage's Expedition 21 (1858).

PLATE XXXVII.

1, two leaves.

2, portion of a leaf.

3, a young flower.

4, a fully developed flower.

5, pistil and stamens.

6, front- and back-view of a stamen.

7, pollen-grain.

8, five fruits.

9, vertical section of a fruit and its calyx.

10, a fruit, the calyx removed, the pistil remaining.

11, a seed.

12, vertical section of a seed.

13, horizontal section of a seed.

All enlarged, but to various extent.

XXXVII

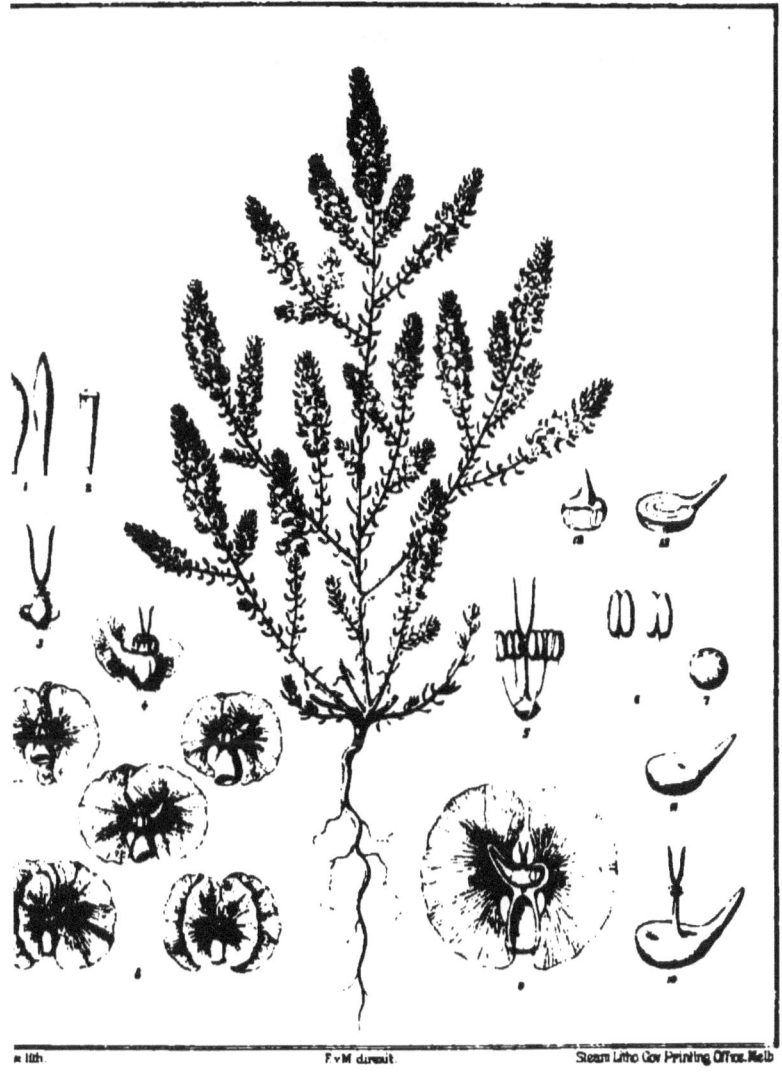

Babbagia dipterocarpa F.v.M.

BABBAGIA SCLEROPTERA.

F. v. M. in Wing's Southern Science Record, November (1885).

PLATE XXXVIII.

1, a leaf and portion of another.

2, a flower.

3, front- and back-view of a stamen.

4, pollen-grain.

5, three fruits with calyx.

6, longitudinal section of a fruit with calyx.

7, a fruit, separated from the calyx, the style and stigmas remaining.

8, longitudinal section of a fruit with style and stigmas.

9, two seeds.

10, horizontal section of a seed.

11, embryo.

All enlarged, but to various extent.

XXXVIII

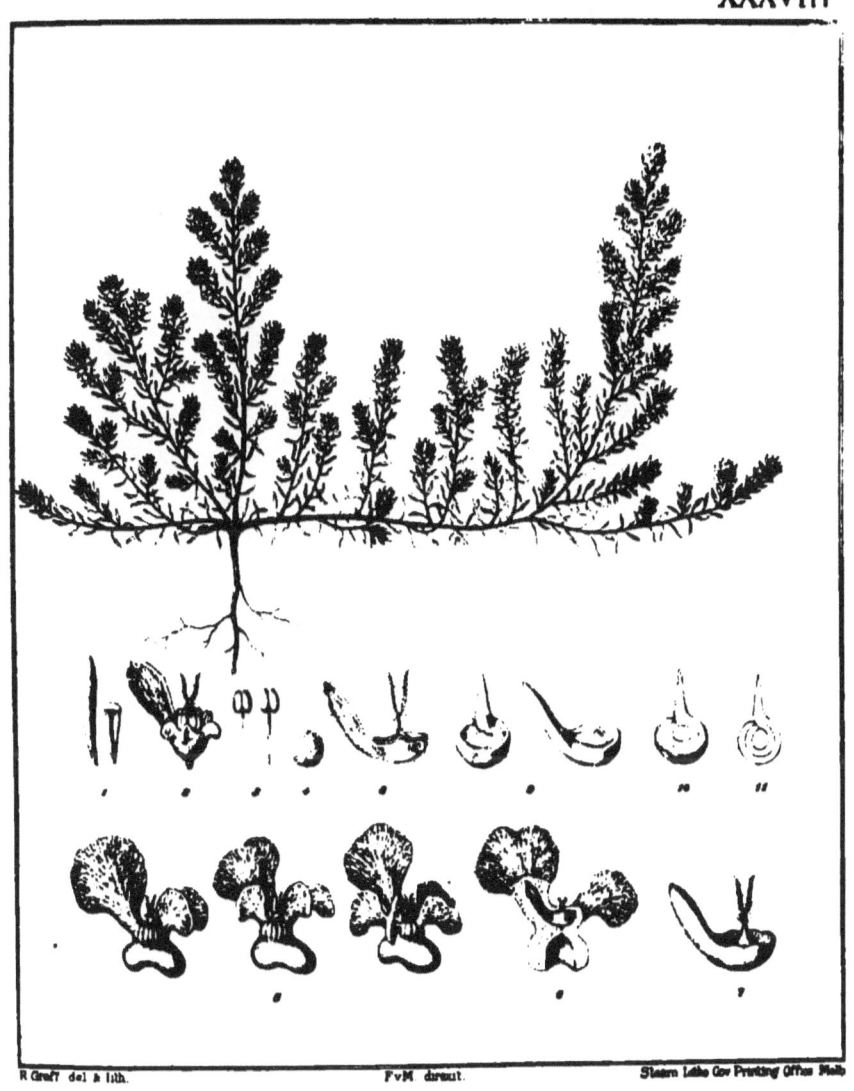

Babbagia scleroptera *FvM*

BABBAGIA ACROPTERA.

F. v. M. and Tate in Transactions of the Royal Society of South Australia vi, 108 (1883).

PLATE XXXIX.

1, a flower.

2, front- and back-view of a stamen.

3, pollen-grain.

4, three fruits with calyx, stamens, style, and stigmas remaining.

5, vertical section of a fruit with calyx, style and stigmas.

6, a fruit separated, with style and stigmas.

7, two seeds.

8, vertical and horizontal section of a seed.

9, embryo.

All enlarged, but to various extent.

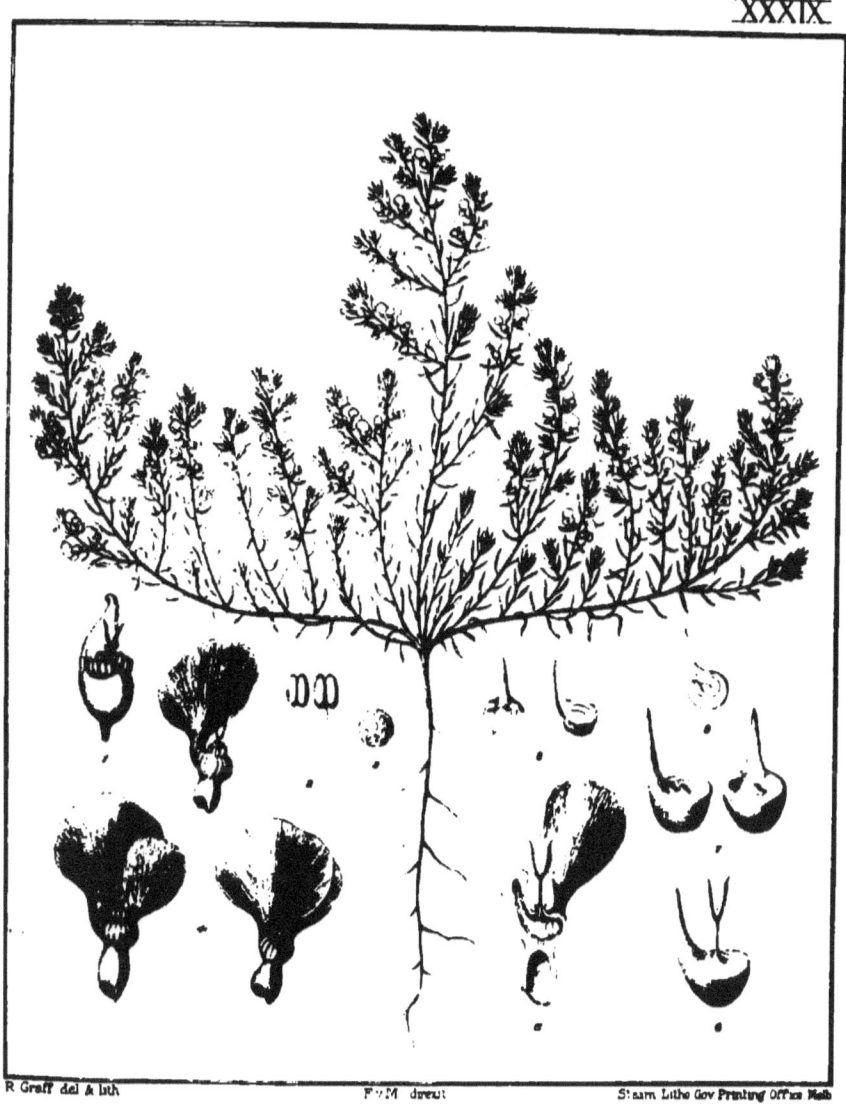

Babbagia acroptera Fv.M & Tate.

Babbagia pentaptera.

F. v. M. and Tate in Transactions of the Royal Society of South Australia vi, 108 (1883).

PLATE XL.

1, a leaf and portion of another.

2, two flowers.

3, front- and back-view of a stamen

4, pollen-grain.

5, portion of a branchlet with a fruit and calyx.

6, a fruit seen from beneath.

7, vertical section of a fruit with calyx, style and stigmas.

8, horizontal section of a fruit with calyx.

9, a fruit separated, with style and stigmas.

10, a seed.

11, embryo.

All enlarged, but to various extent.

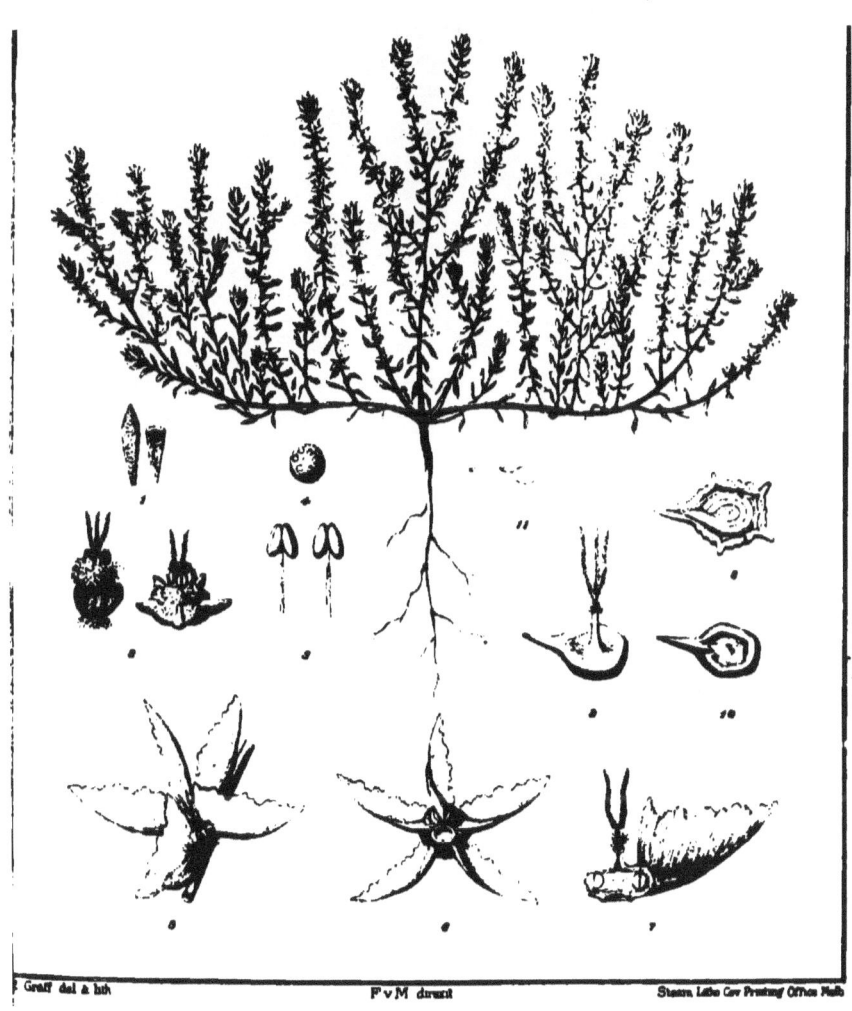

Babbagia pentaptera F.v.M. & TATE.

ICONOGRAPHY

OF

AUSTRALIAN SALSOLACEOUS PLANTS.

BY

BARON FERD. VON MUELLER, K.C.M.G., M. & PH.D., F.R.S.,

GOVERNMENT BOTANIST OF THE COLONY OF VICTORIA.

FIFTH DECADE.

By Authority:
ROBERT S. BRAIN, GOVERNMENT PRINTER, MELBOURNE.
1890.

KOCHIA DICHOPTERA.

F. v. Mueller, fragmenta phytographiae Australiae viii, 37 (1873).

PLATE XLI.

1, portions of leaves.
2, a flower.
3, front- and back-view of a stamen.
4, pollen-grain.
5, fruit-bearing calyces, two seen from above, one from beneath.
6, fruit-bearing calyces still further enlarged.
7, vertical section of a fruit with its calyx.
8, a fruit, the calyx removed.
9, a seed.
10, horizontal section of a seed.

All enlarged, but to various extent.

XLI

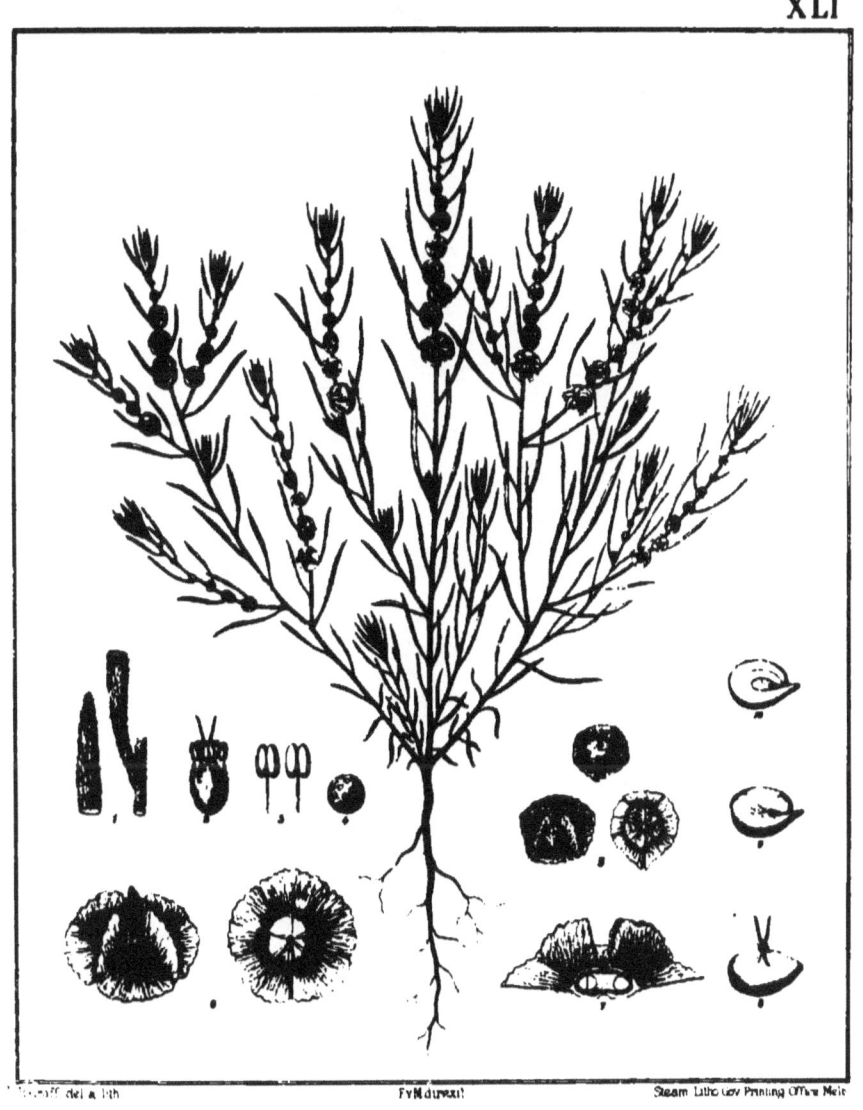

Kochia dichoptera F.v.M.

KOCHIA OPPOSITIFOLIA.

F. v. Mueller in Transactions of the Victorian Institute 134 (1855).

PLATE XLII.

1, portion of a branchlet with leaves and flower-buds.

2, two flowers.

3, front- and back-view of a stamen.

4, pollen-grain.

5, fruit-bearing calyces, variously viewed.

6, vertical section of a fruit with its calyx.

7, two fruits, the calyx removed.

8, a seed.

9, horizontal section of a seed.

All enlarged, but to various extent.

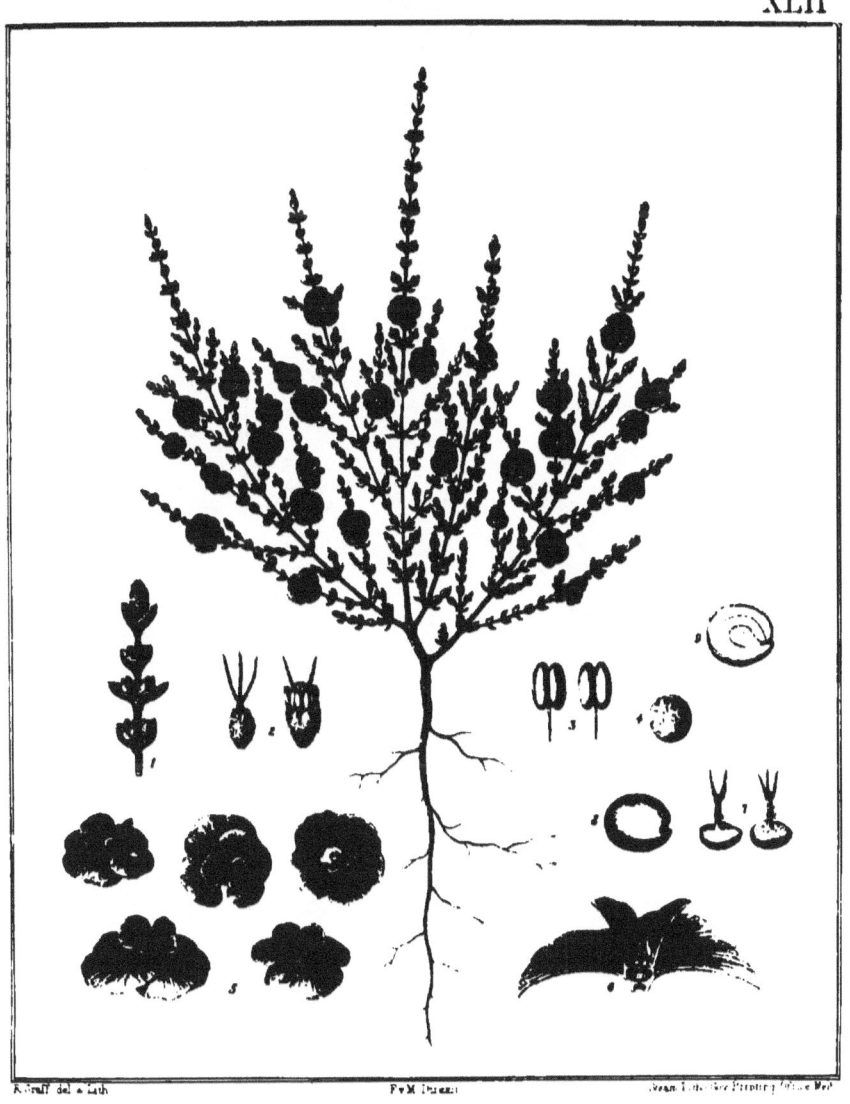

Kochia oppositifolia *FvM.*

KOCHIA BREVIFOLIA.

R. Brown, prodromus florae Novae Hollandiae 409 (1810).

PLATE XLIII.

1, portion of two branchlets with leaves and flowers.

2, two flowers in different stages of development.

3, front- and back-view of a stamen.

4, pollen-grain.

5, two fruit-bearing calyces, one seen from above, the other from beneath.

6, vertical section of a fruit with its calyx.

7, a fruit, the calyx removed.

8, a seed.

9, horizontal section of a seed.

All enlarged, but to various extent.

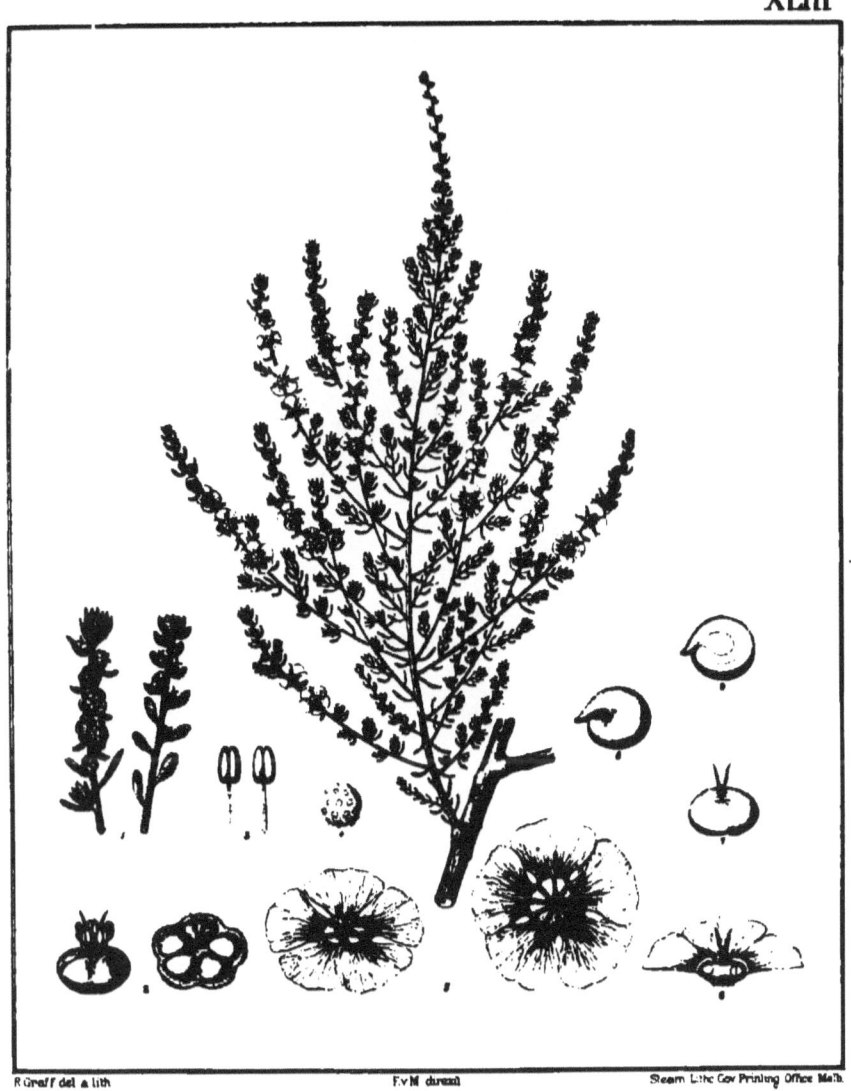

Kochia brevifolia R. Brown.

KOCHIA FIMBRIOLATA.

F. v. Mueller, fragmenta phytographiae Australiae ix, 75 (1875).

PLATE XLIV.

1, portion of a branchlet with leaves and fruits.
2, a flower.
3, vertical section of a flower.
4, front- and back-view of a stamen.
5, pollen-grain.
6, side-view of a fruit-bearing calyx.
7, back-view of a fruit-bearing calyx.
8, vertical section of a fruit with its calyx.
9, a fruit, the calyx removed.
10, a seed.
11, horizontal section of a seed.

All enlarged, but to various extent.

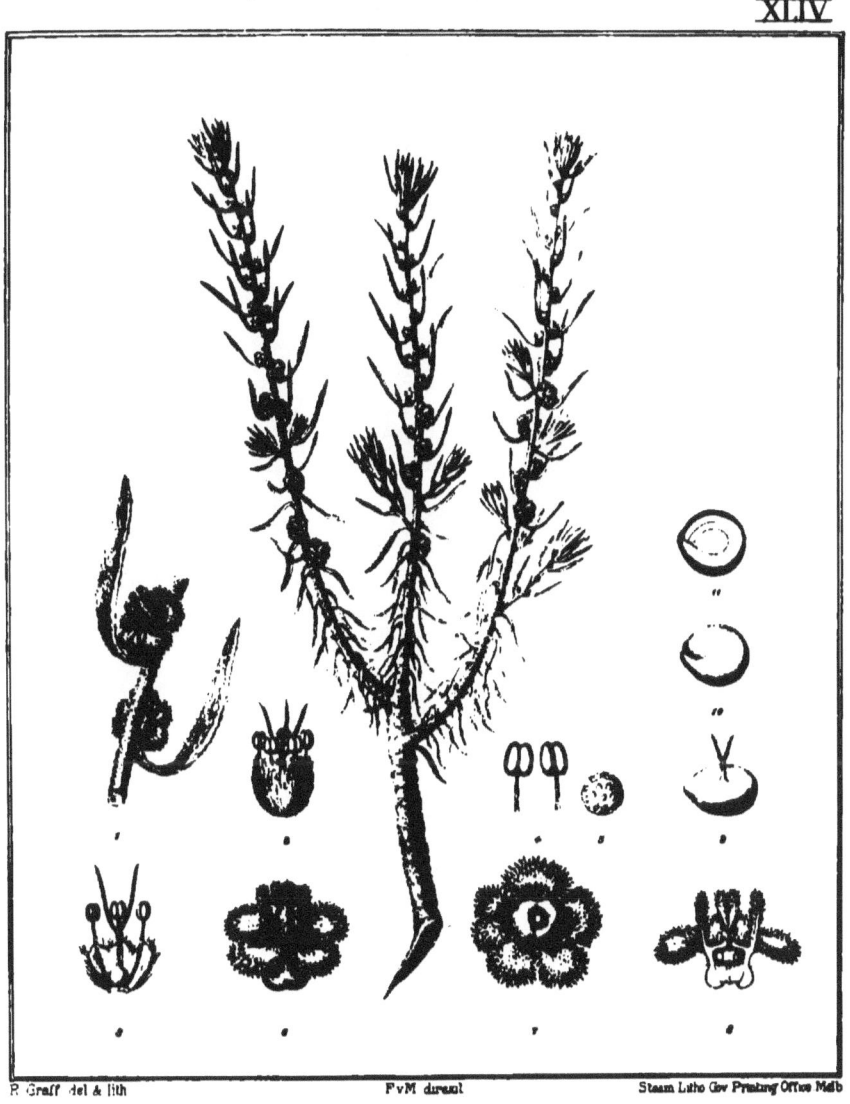

Kochia fimbriolata FvM.

KOCHIA LOBIFLORA.

F. v. Mueller in Bentham's Flora Australiensis v, 184 (1870).

PLATE XLV.

1, portion of a branchlet with leaves and flowers.

2, front- and back-view of a stamen.

3, pollen-grain.

4, side-view of a fruit-bearing calyx.

5, vertical section of two fruits with their calyces.

6, back-view of a fruit with its calyx.

7, a fruit, the calyx removed.

8, a seed.

9, horizontal section of a seed.

All enlarged, but to various extent.

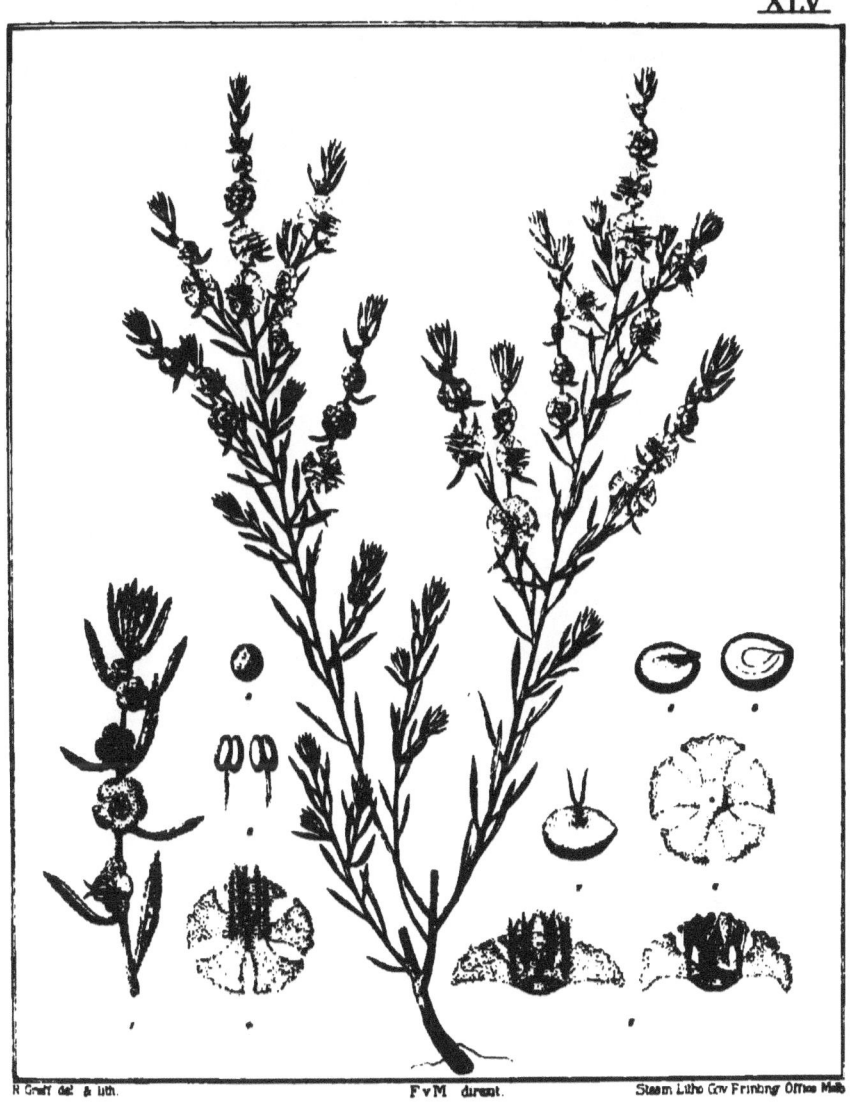

Kochia lobiflora *F.v.M.*

KOCHIA LANOSA.

Lindley in Mitchell's Tropical Australia 88 (1848).

PLATE XLVI.

1, portion of two leaves.

2, a flower.

3, front- and back-view of a stamen.

4, pollen-grain.

5, side-view of a fruit-bearing calyx.

6, back-view of a fruit-bearing calyx.

7, vertical section of a fruit with its calyx.

8, a fruit, the calyx removed.

9, a seed.

10, horizontal section of a seed.

All enlarged, but to various extent.

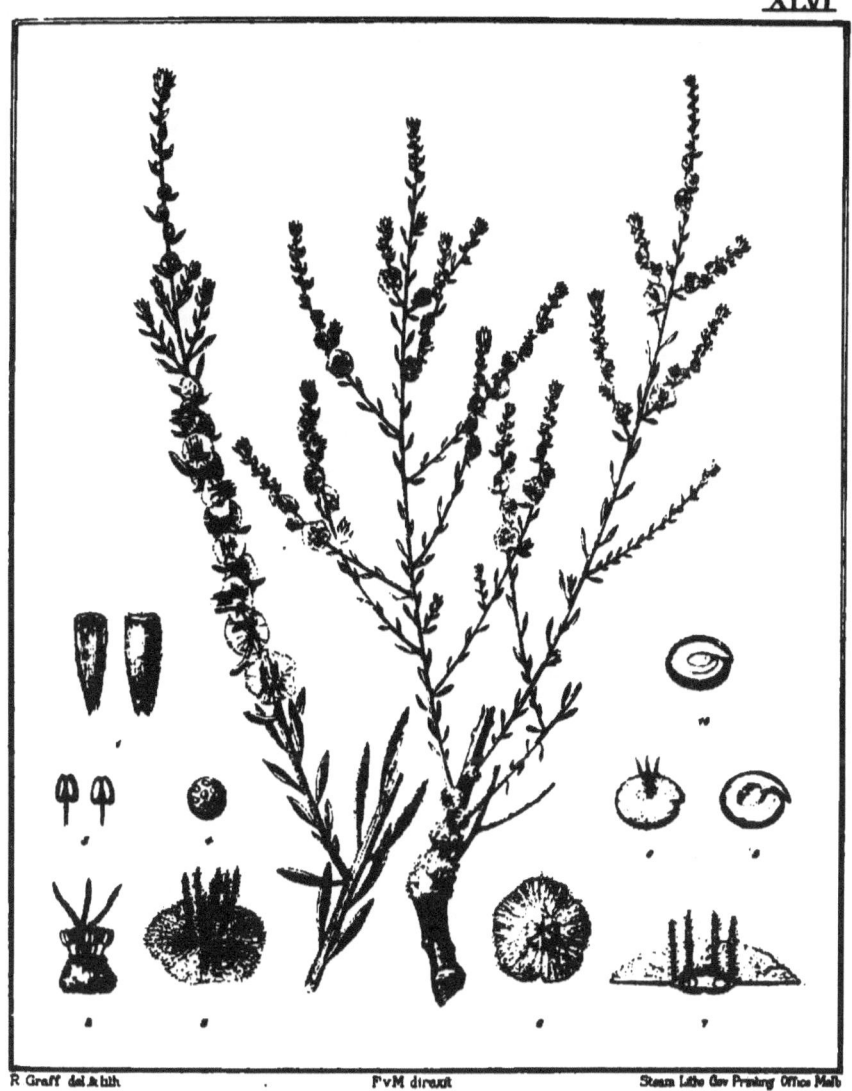

Kochia lanosa *Lindley*.

KOCHIA PROSTECOCHAETA.

F. v. Mueller, fragmenta phytographiae Australiae xii, 14 (1882).

PLATE XLVII.

1, portions of leaves.
2, a flower.
3, front- and back-view of a stamen.
4, pollen-grain.
5, a young fruit-bearing calyx.
6, a mature fruit-bearing calyx.
7, vertical section of a fruit with its calyx.
8, a fruit, the calyx removed.
9, a seed.
10, vertical section of a seed.
11, horizontal section of a seed.

All enlarged, but to various extent.

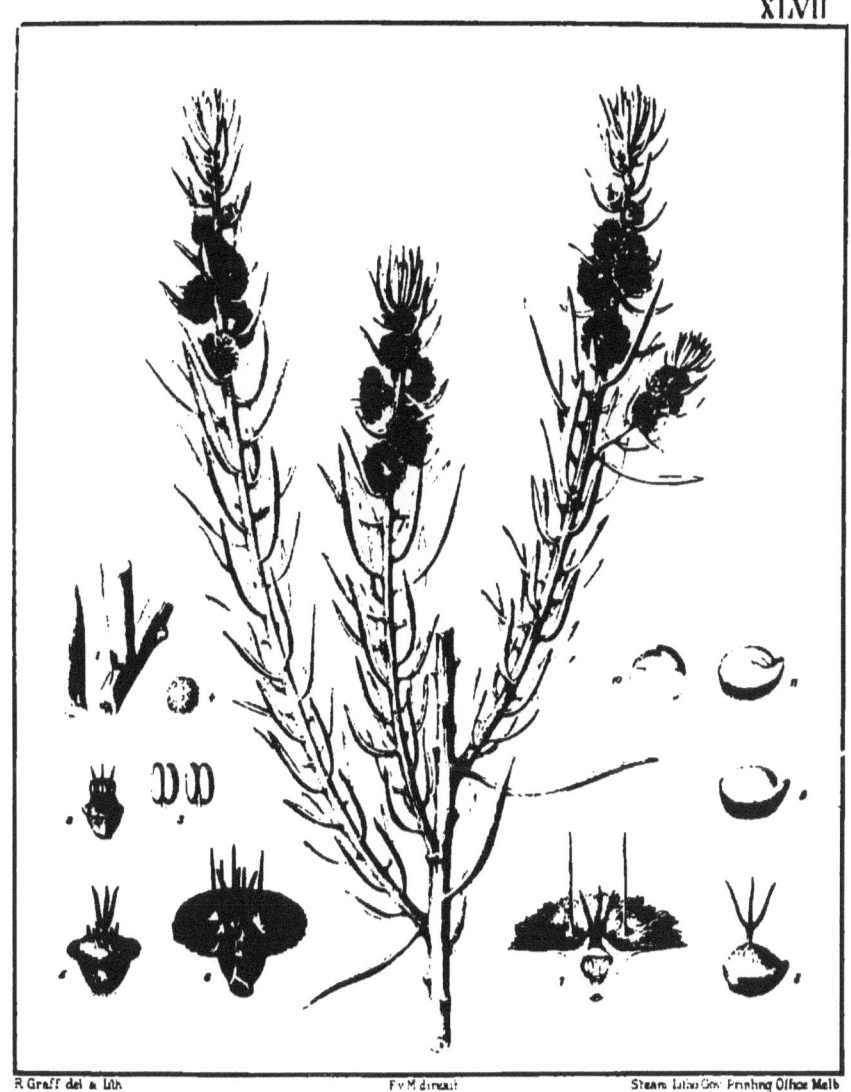

Kochia prosthecochaeta FvM.

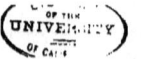

KOCHIA MELANOCOMA.

F. v. Mueller, fragmenta phytographiae Australiae xii, 14 (1882).

PLATE XLVIII.

1, portion of two leaves.

2, a flower.

3, vertical section of a flower.

4, front- and back-view of a stamen.

5, pollen-grain.

6, a young fruit-bearing calyx.

7, a mature fruit-bearing calyx.

8, vertical section of a fruit with its calyx.

9, a fruit, the calyx removed.

10, a seed.

11, horizontal section of a seed.

All enlarged, but to various extent.

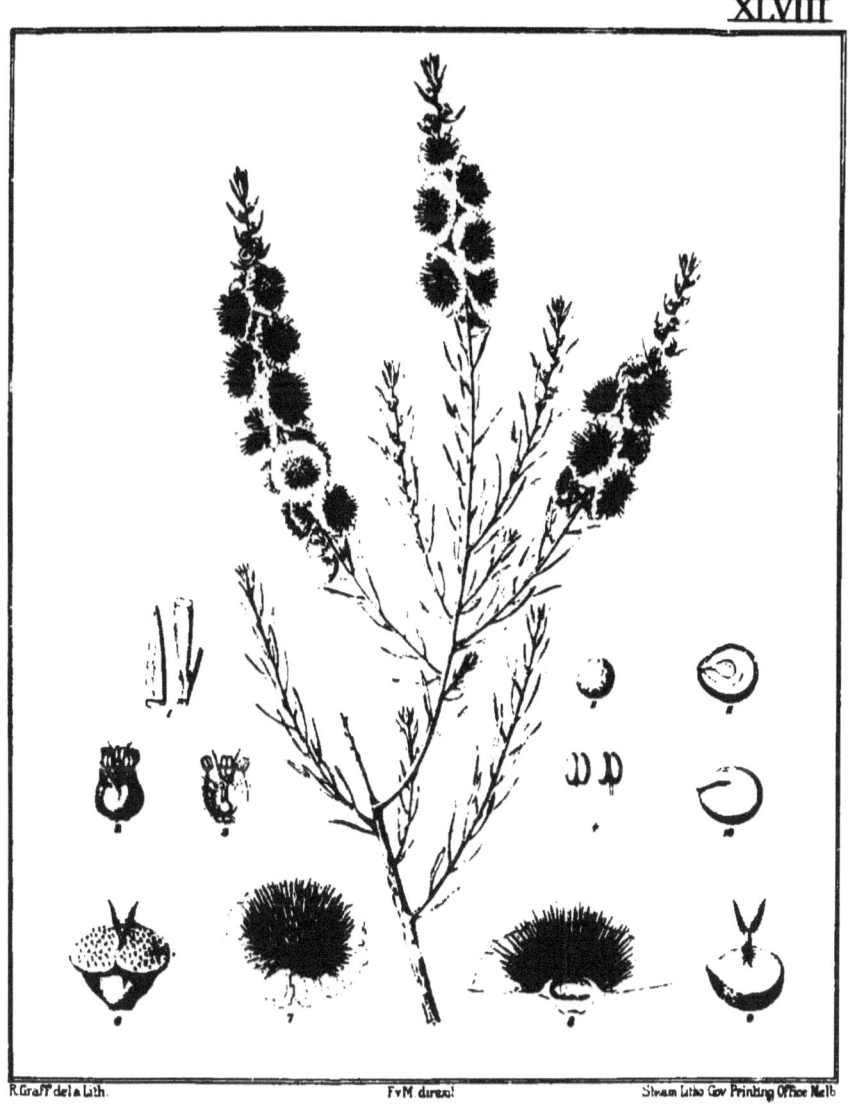

Kochia melanocoma F.v.M.

KOCHIA PYRAMIDATA.

Bentham, Flora Australiensis v, 186 (1870).

PLATE XLIX.

1 and 2, portion of two branchlets with leaves and flowers.

3, portion of two leaves.

4, two flowers.

5, a flower in a more advanced state.

6, front- and back-view of a stamen.

7, pollen-grain.

8, two flowers far advanced, one with half the calyx removed.

9, two fruit-bearing calyces.

10, vertical section of a fruit with its calyx.

11, a fruit, the calyx removed.

12, a seed.

13, horizontal section of a seed.

All enlarged, but to various extent.

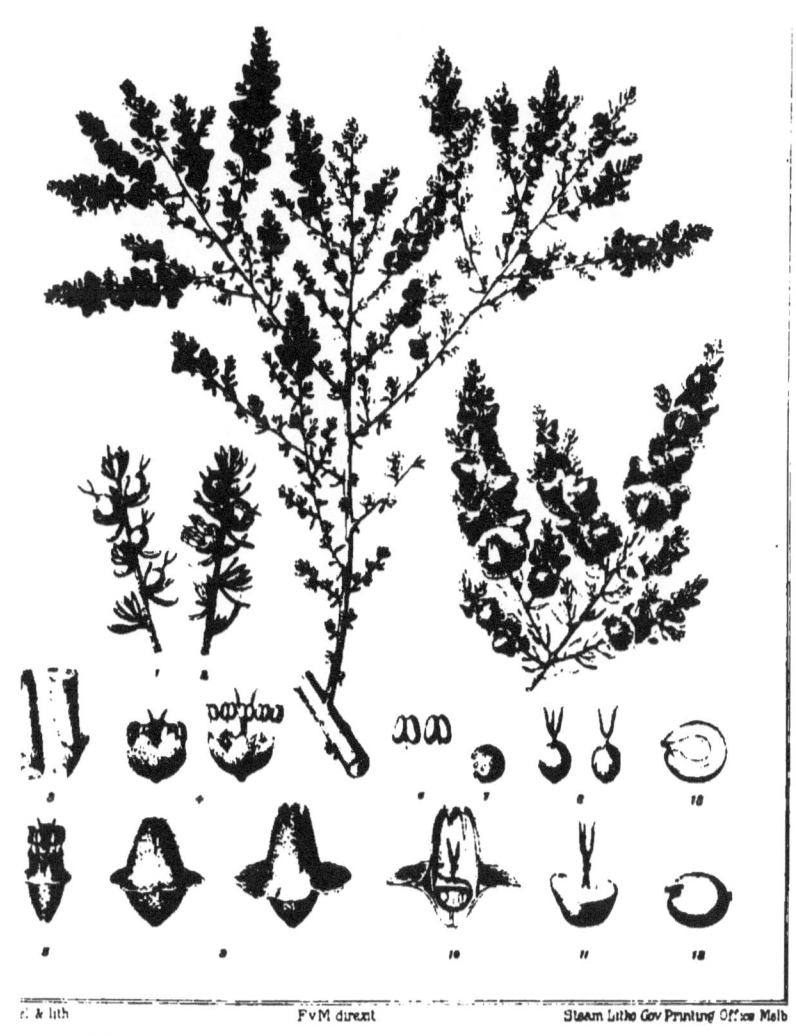

Kochia pyramidata Bentham.

KOCHIA TRIPTERA.

Bentham, Flora Australiensis v, 185 (1870).

PLATE L.

1, portion of two leaves.

2, two flowers.

3, front- and back-view of a stamen.

4, pollen-grain.

5, fruit-bearing calyces, variously viewed.

6, vertical section of a fruit with its calyx.

7, two fruits, the calyx removed.

8, a seed.

9, horizontal section of a seed.

All enlarged, but to various extent.

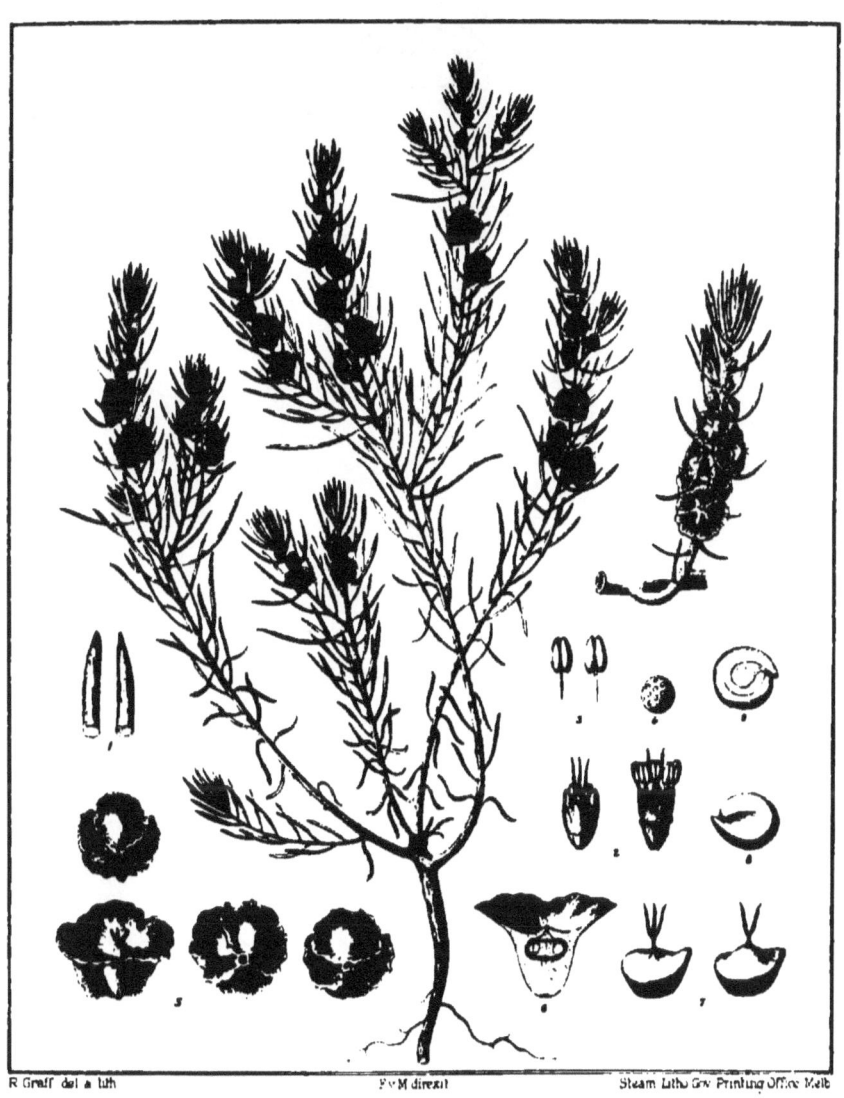

Kochia triptera Bentham

WILLIAM A. SETCHELL,
UNIV. OF CALIFORNIA,
BERKELEY, - - - CALIF.

ICONOGRAPHY

OF

AUSTRALIAN SALSOLACEOUS PLANTS

BY

BARON FERD. VON MUELLER, K.C.M.G., M. & PH.D., F.R.S.,

GOVERNMENT BOTANIST OF THE COLONY OF VICTORIA.

"MULTI SUNT MUTI VOCATI, PAUCI VERO ELECTI."—*Evang. Matth. xxii., 14.*

SIXTH DECADE.

By Authority:
ROBERT S. BRAIN, GOVERNMENT PRINTER, MELBOURNE.
1890.

KOCHIA SPONGIOCARPA.

F. v. Mueller in Victorian Naturalist iii, 92 (1886).

PLATE LI.

1, portion of two leaves.
2, a flower.
3, back- and front-view of a stamen.
4, pollen-grain.
5, side-view of a fruit-bearing calyx.
6, back-view of a fruit-bearing calyx.
7, vertical section of a fruit with its calyx.
8, a fruit, the calyx removed.
9, a seed.
10, vertical section of a seed.
11, horizontal section of a seed.

All enlarged, but to various extent.

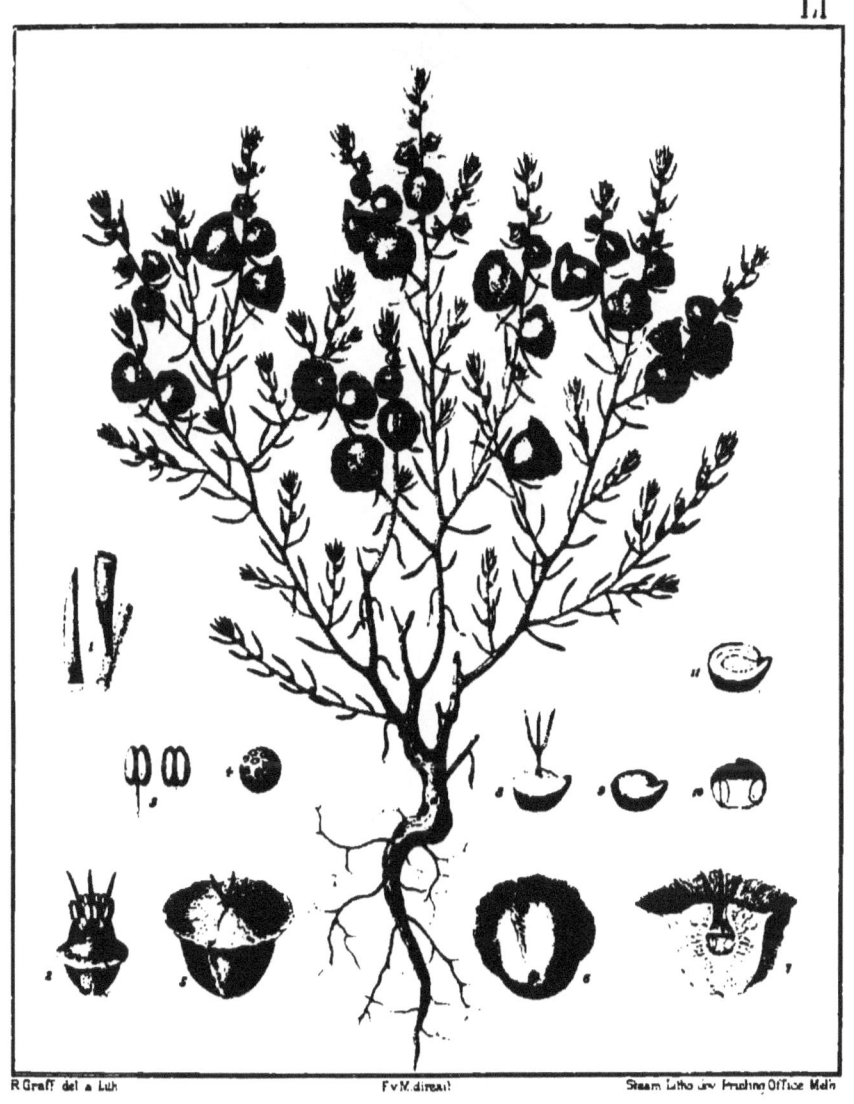

Kochia spongiocarpa *F.v.M.*

KOCHIA MICROPHYLLA.

F. v. Mueller, fragmenta phytographiae Australiae viii, 148 (1874).

PLATE LII.

1, portion of two branchlets with leaves and flowers.

2, a flower.

3, front- and back-view of a stamen.

4, pollen-grain.

5, a fruit-bearing calyx, seen from above.

6, vertical section of a fruit with its calyx.

7, a fruit, the calyx removed.

8, vertical section of a fruit without its calyx.

9, a seed.

10, horizontal section of a seed.

All enlarged, but to various extent

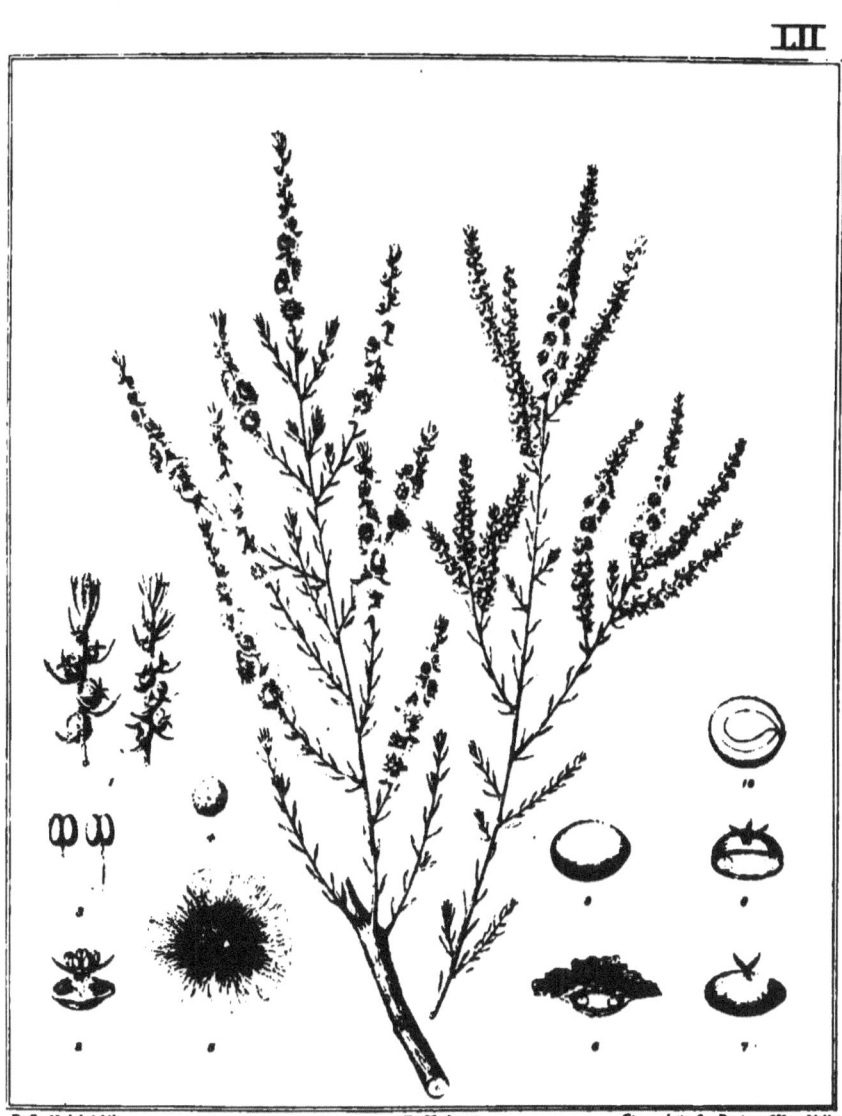

Kochia microphylla *FvM*

KOCHIA VILLOSA.

Lindley in Mitchell's Tropical Australia 91 (1848).

PLATE LIII.

1, portion of three leaves.
2, three flowers in different stages.
3, front- and back-view of a stamen.
4, pollen-grain.
5, three fruit-bearing calyces, two seen from above, one from below.
6, vertical section of fruits with their calyces.
7, two fruits, the calyx removed.
8, a seed.
9, horizontal section of a seed.

All enlarged, but to various extent.

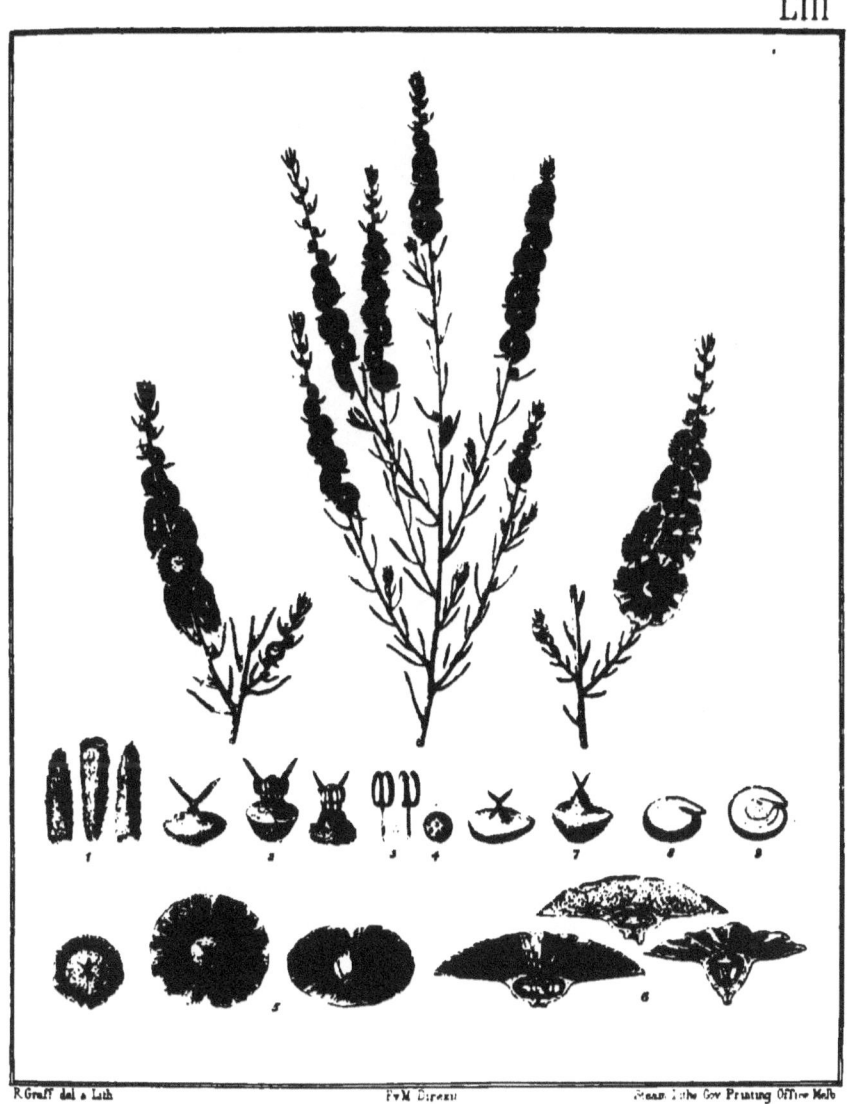

Kochia villosa *Lindley*

KOCHIA SEDIFOLIA.

F. v. Mueller in Transactions of the Victorian Institute i, 134 (1855).

PLATE LIV.

1, leaves in various positions.

2, a flower.

3, front- and back-view of a stamen.

4, pollen-grain.

5, three flowers, advanced.

6, a flower still more advanced.

7, front-view of a fruit-bearing calyx.

8, back-view of a fruit-bearing calyx.

9, vertical section of a fruit with its calyx.

10, two fruits, the calyx removed.

11, a seed.

12, horizontal section of a seed.

All enlarged, but to various extent.

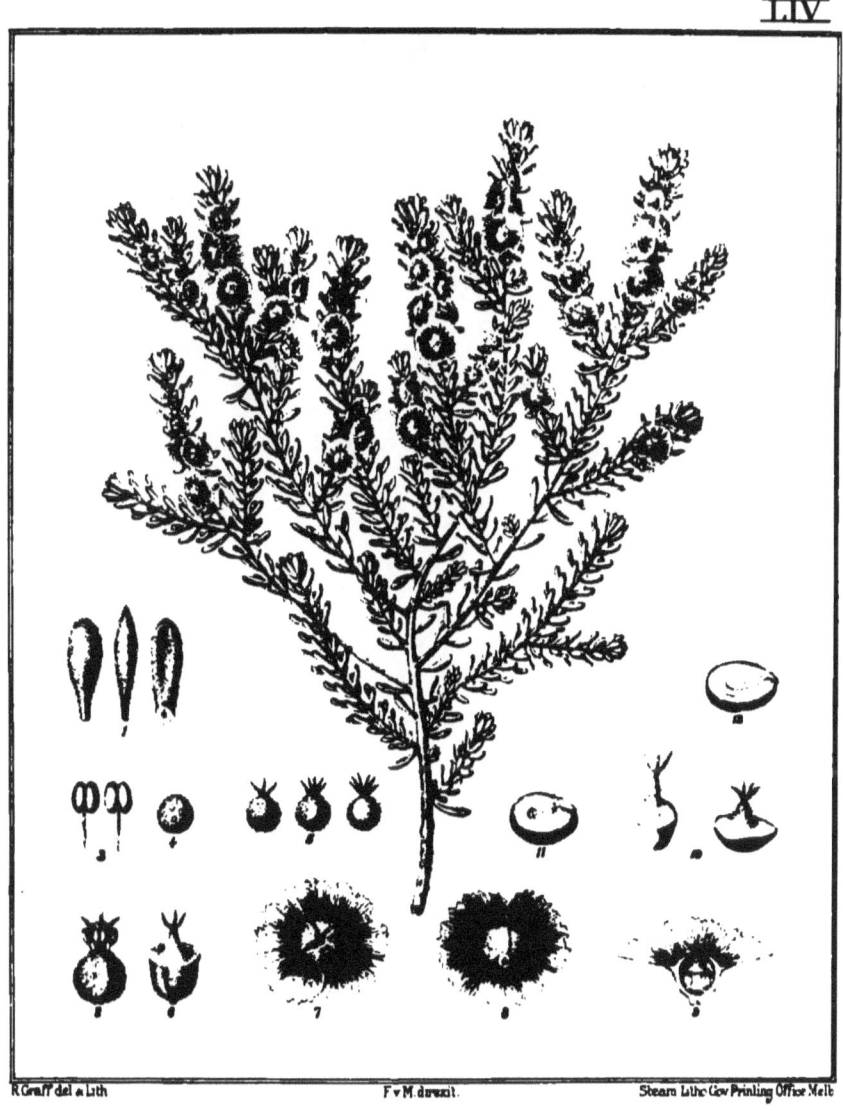

Kochia sedifolia F.v.M.

KOCHIA APHYLLA.

R. Brown, prodromus florae Novae Hollandiae 409 (1810).

PLATE LV.

1, leaves in various positions.

2, a flower.

3, front- and back-view of a stamen.

4, pollen-grain.

5, vertical section of a young fruit with portion of its calyx.

6, fruit-bearing calyces, one seen from above, the other from below.

7, vertical section of a fruit with its calyx.

8, a fruit, the calyx removed.

9, a seed.

10, horizontal section of a seed.

11, section of a gall.

All enlarged, but to various extent.

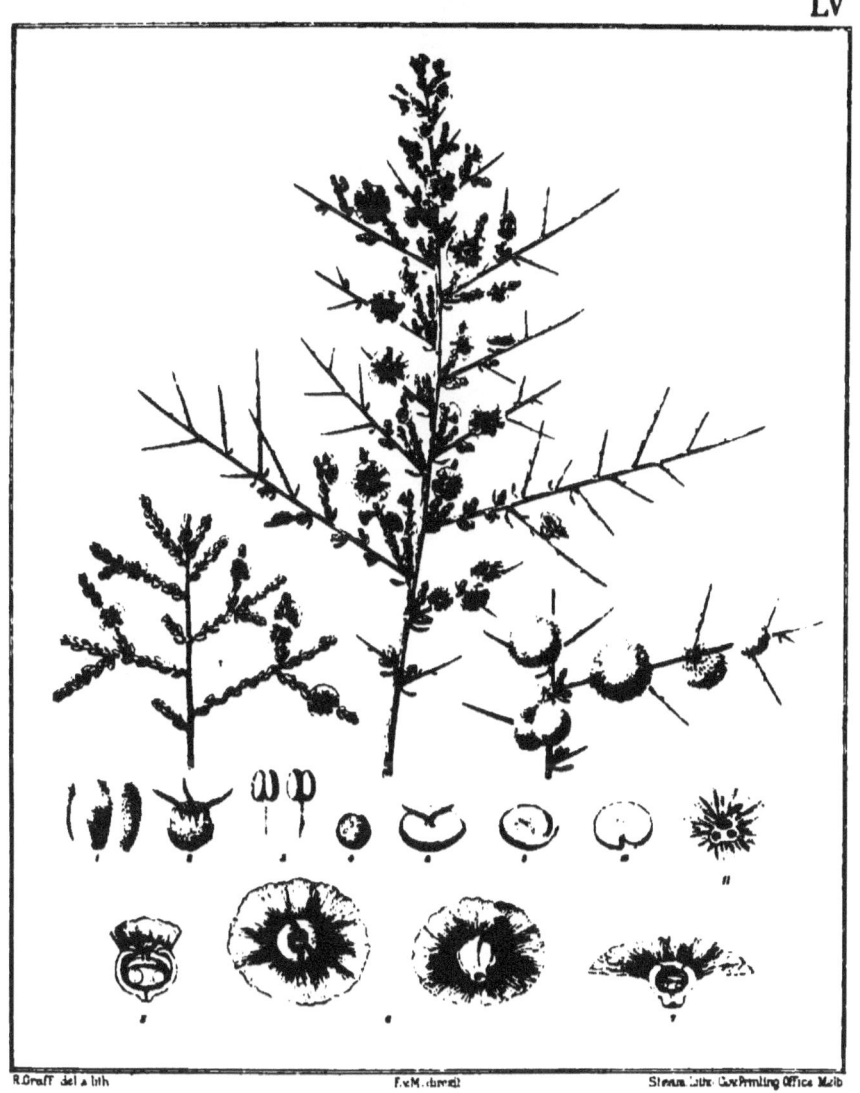

Kochia aphylla R. BROWN.

Kochia humillima.

F. v. Mueller, fragmenta phytographiae Australiae ix, 168 (1875).

PLATE LVI.

1, portion of a young leafy shoot.

2, portion of a branchlet with one developed and several young leaves.

3, two flowers.

4, three pistils with various stigmas.

5, front- and back-view of a stamen.

6, pollen-grain.

7, two fruit-bearing calyces, seen from above.

8, vertical section of a fruit with its calyx.

9, a fruit, the calyx removed.

10, a seed.

11, vertical section of a seed.

12, horizontal section of a seed.

All enlarged, but to various extent.

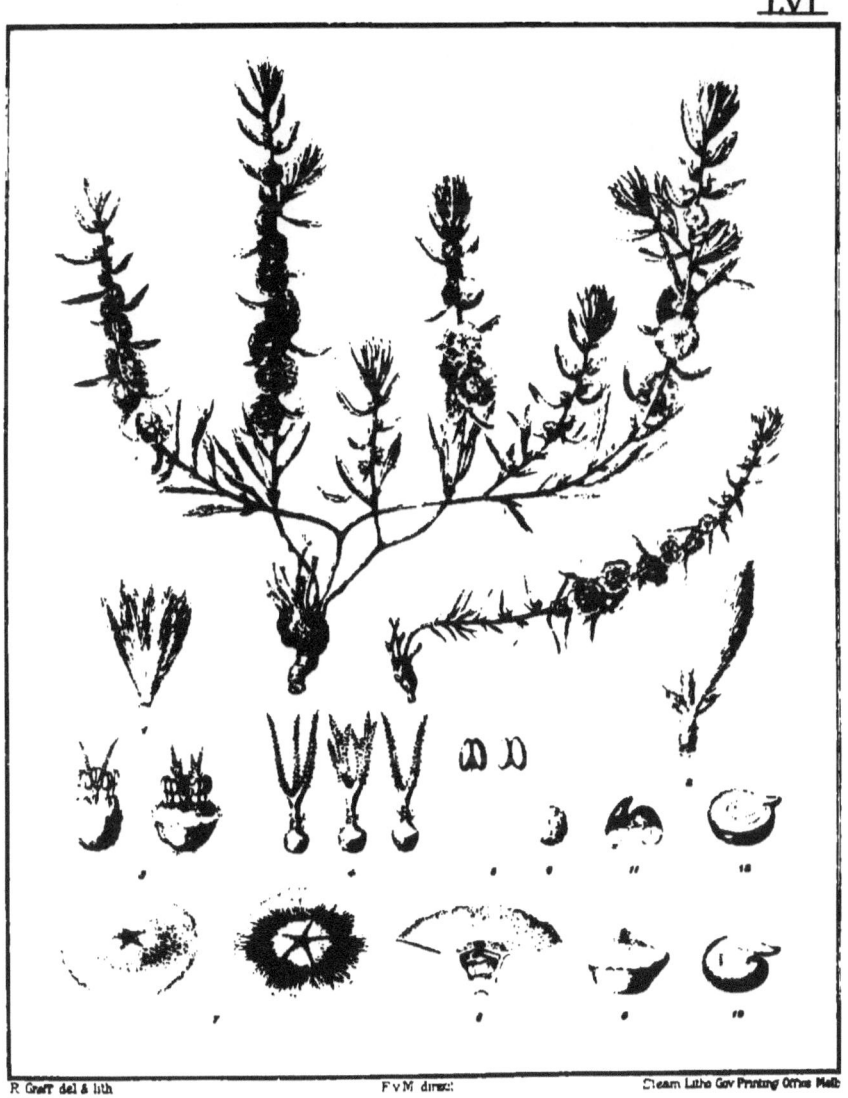

Kochia humillima *FvM*

KOCHIA ERIANTHA.

F. v. Mueller, Report on the Plants of Babbage's Expedition 20 (1858).

PLATE LVII.

1, portion of two leaves.

2, two flowers.

3, front- and back-view of a stamen.

4, pollen-grain.

5, side-view of a calyx in an advanced state.

6, a fruit-bearing calyx, seen from above.

7, vertical section of a fruit with its calyx.

8, a fruit, the calyx removed.

9, a seed.

10, horizontal section of a seed.

All enlarged, but to various extent.

Kochia eriantha F.v.M.

KOCHIA CILIATA.

F. v. Mueller, Report on Plants of Babbage's Expedition 20 (1858).

PLATE LVIII.

1, portion of a leaf.

2, portion of a branchlet with leaves and flowers.

3, two flowers.

4, front- and back-view of a stamen.

5, pollen-grain.

6, a fruit-bearing calyx.

7, vertical section of two fruits with their calyces.

8, a fruit, the calyx removed.

9, a seed.

10, horizontal section of a seed.

All enlarged, but to various extent.

LVIII

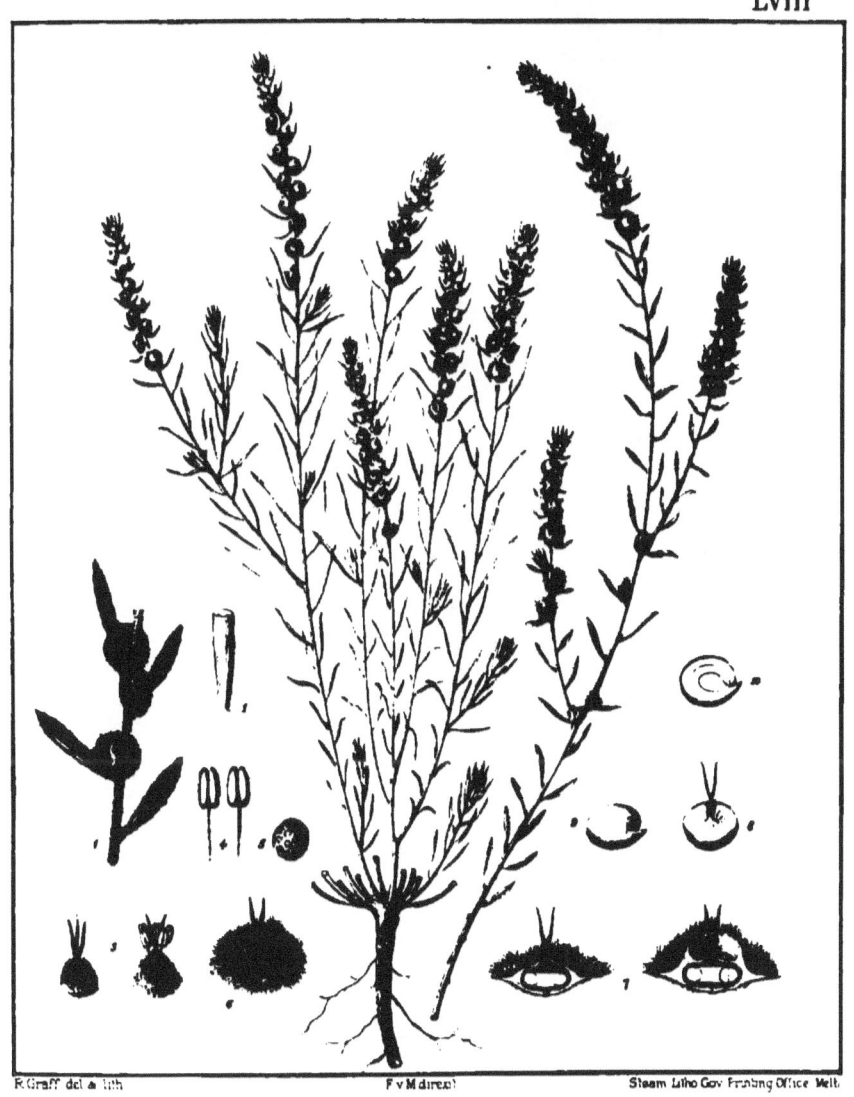

Kochia ciliata F.v.M.

KOCHIA BRACHYPTERA.

F. v. Mueller, Second General Report 15 (1854).

PLATE LIX.

1, portion of two leaves.
2, a flower.
3, front- and back-view of a stamen.
4, pollen-grain.
5, two fruit-bearing calyces, side-view.
6, two fruit-bearing calyces, back-view.
7, vertical section of a fruit with its calyx.
8, a fruit, the calyx removed.
9, a seed.
10, horizontal section of a seed.

All enlarged, but to various extent.

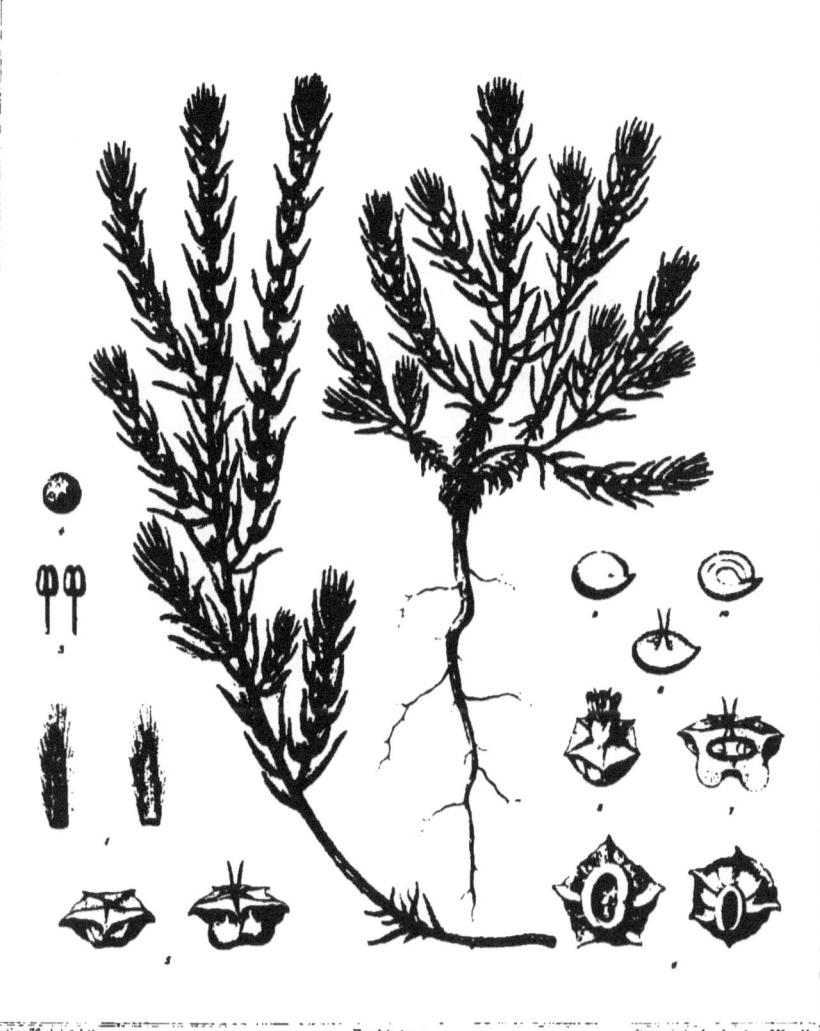

Kochia brachyptera *FvM*

DIDYMANTHUS ROEI.

Endlicher, Novarum Stirpium Decades 8 (1839)

PLATE LX.

1, portion of two leaves.

2, twin flowers.

3, twin flowers, the calyx of one partly removed.

4, front- and back-view of a stamen.

5, pollen-grain.

6 and 7, two pairs of fruit-bearing calyces.

8, vertical section of a pair of fruits with their calyces.

9, a fruit, the calyx removed.

10, a seed.

11, transverse section of a fruit.

12, longitudinal section of a seed.

All enlarged, but to various extent.

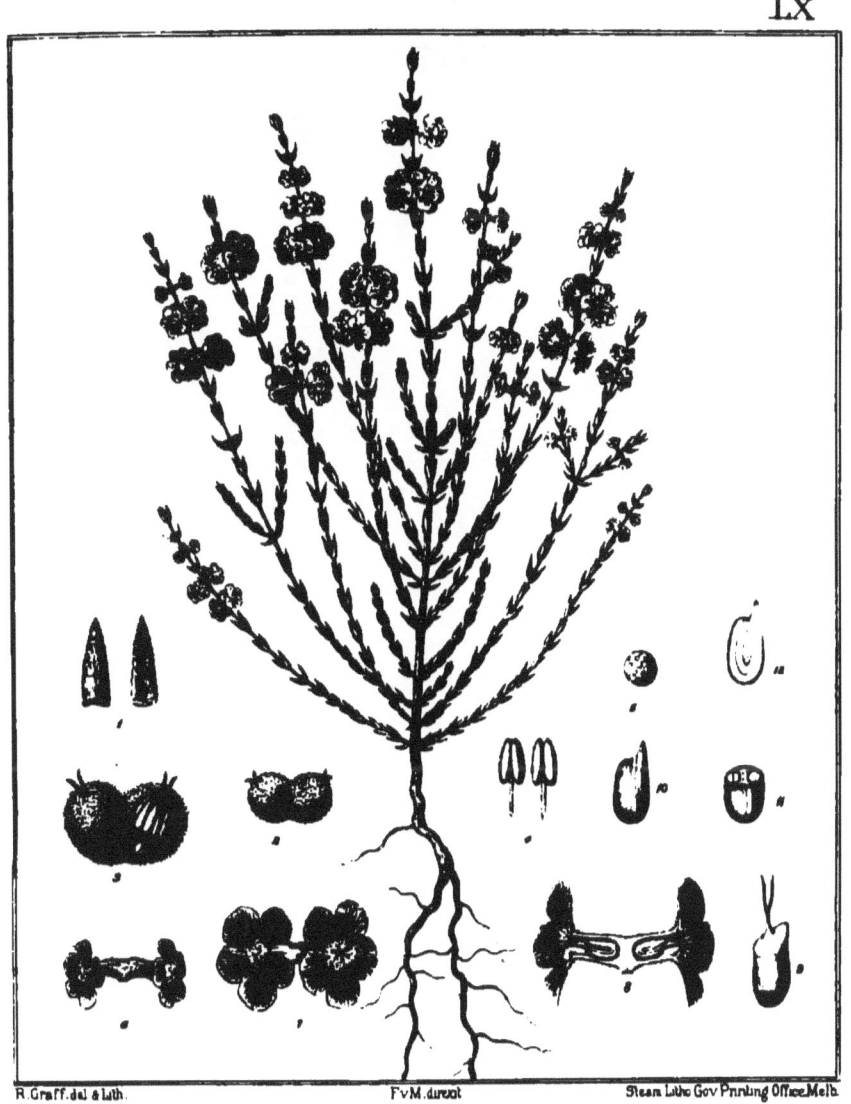

Didymanthus Roei *Endlicher*

WILLIAM A. SETCHEL..
UNIV OF CALIFORNIA.
BERKELEY, - - - CALIF.

ICONOGRAPHY

OF

AUSTRALIAN SALSOLACEOUS PLANTS

BY

BARON FERD. VON MUELLER, K.C.M.G., M. & PH.D., F.R.S.,

GOVERNMENT BOTANIST OF THE COLONY OF VICTORIA.

"Moras huic novi vocati, paroi viro selecti."—*Plumy. Nostk. xxii., 14.*

SEVENTH DECADE.

By Authority:
ROBT. S. BRAIN, GOVERNMENT PRINTER, MELBOURNE.
1891.

BASSIA BIFLORA.

F. v. M., Census of Australian Plants 30 (1882).

PLATE LXI.

1, portions of leaves.

2, a pair of flowers.

3, front- and back-view of a stamen.

4, pollen-grain.

5, different connate fruits with calyces.

6, longitudinal section of connate fruits with calyces.

7, two fruits, the calyx removed, the style and stigmas remaining.

8, a seed.

9, longitudinal section of a seed.

All enlarged, but to various extent.

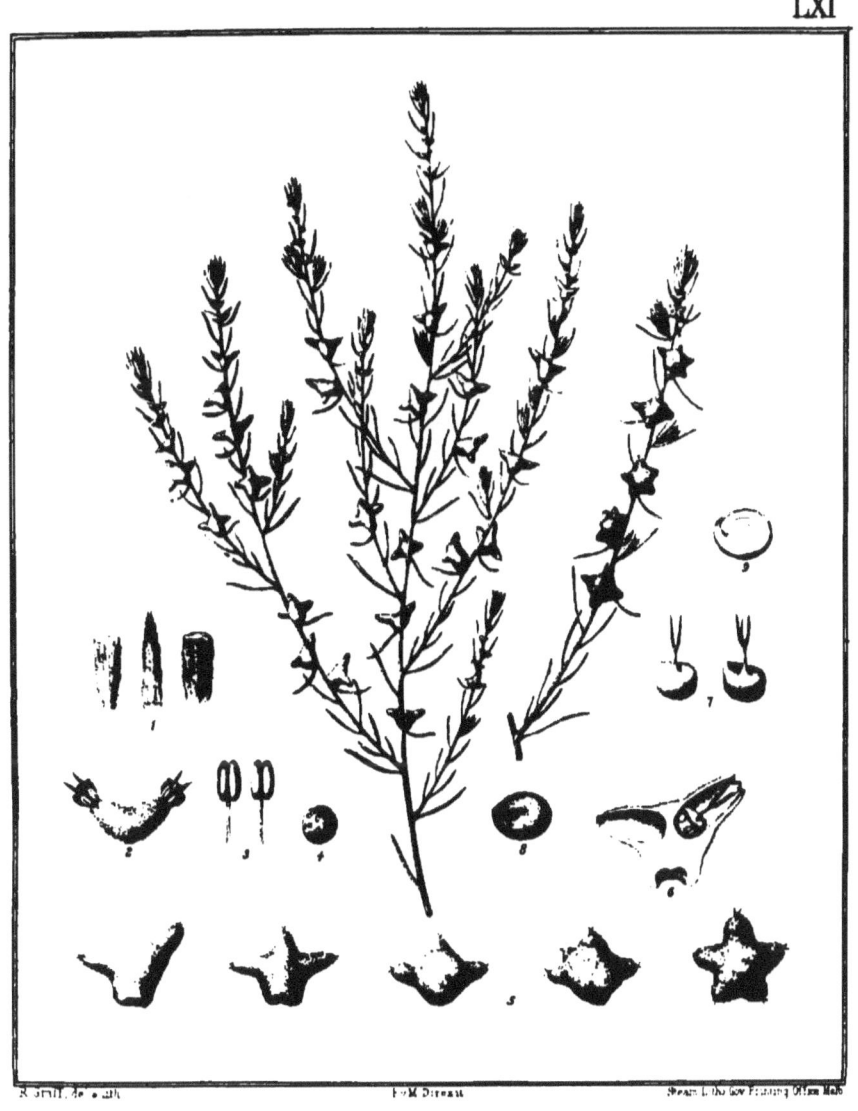

Bassia biflora *FvM*

Bassia paradoxa.

F. v. M., Census of Australian Plants 30 (1882).

PLATE LXII.

1, portions of leaves.
2, a lot of connate flowers.
3, front- and back-view of a stamen.
4, pollen-grain.
5, vertical section of a lot of fruits with their connate calyces.
6, a fruit, the calyx removed, the style and stigmas remaining.
7, two seeds.
8, longitudinal section of a seed.

All enlarged, but to various extent.

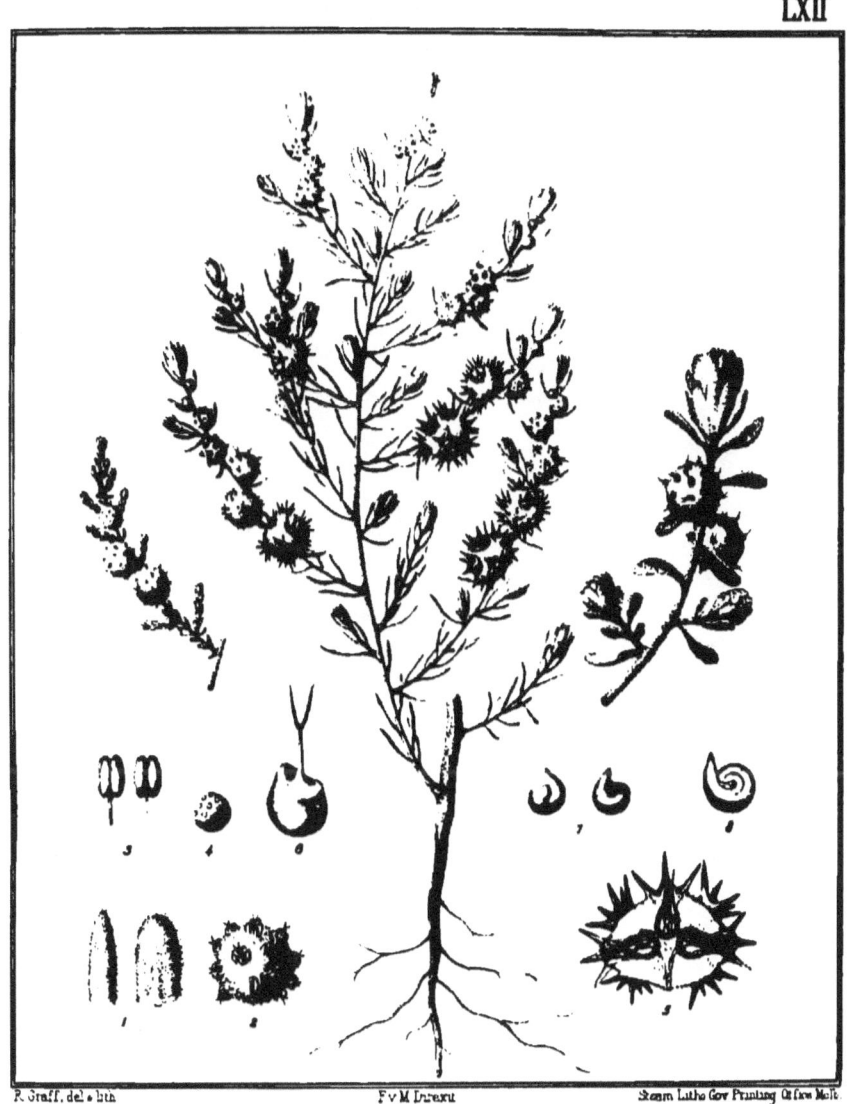

Bassia paradoxa F.v.M.

BASSIA TRICORNIS.

F. v. M., Census of Australian Plants 30 (1882).

PLATE LXIII.

1, portions of leaves.
2, a flower.
3, front- and back-view of a stamen.
4, pollen-grain.
5, fruits with their calyces.
6, vertical section of a fruit with calyx.
7, fruits, the calyx removed, the style and stigma remaining.
8, a seed.
9, transverse section of a seed.
10, longitudinal section of a seed.

      All enlarged, but to various extent.

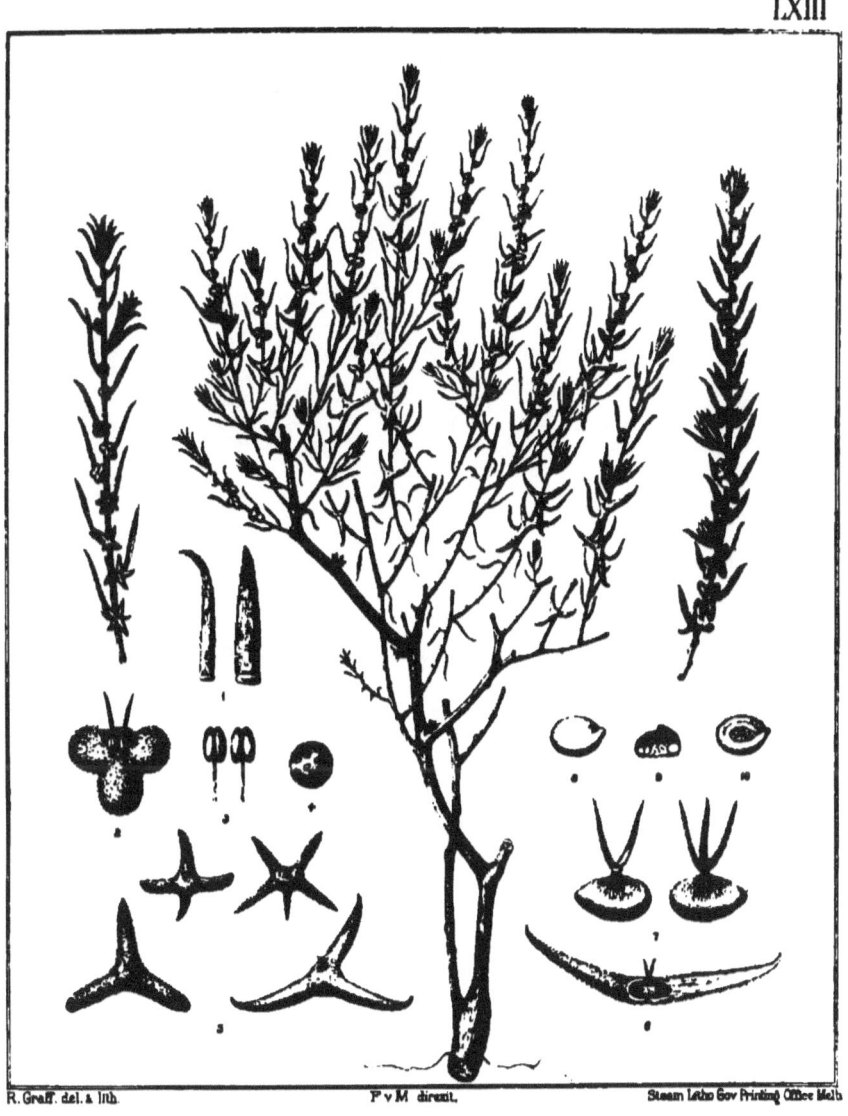

Bassia tricornis F.v.M.

BASSIA TRIDENS.

F. v. M., fragmenta phytographiae Australiae xii, 12 (1882).

PLATE LXIV.

1, branchlet with flowers.
2, portions of leaves.
3, flowers in different stages of development.
4, front- and back-view of a stamen.
5, pollen-grain.
6, a young fruit with calyx, style and stigmas.
7, a developed fruit with calyx, style and stigma.
8, vertical section of a fruit with calyx, style and stigmas.
9, a fruit, the calyx removed, the style and stigmas remaining.
10, a seed.
11, longitudinal section of a seed.

All enlarged, but to various extent.

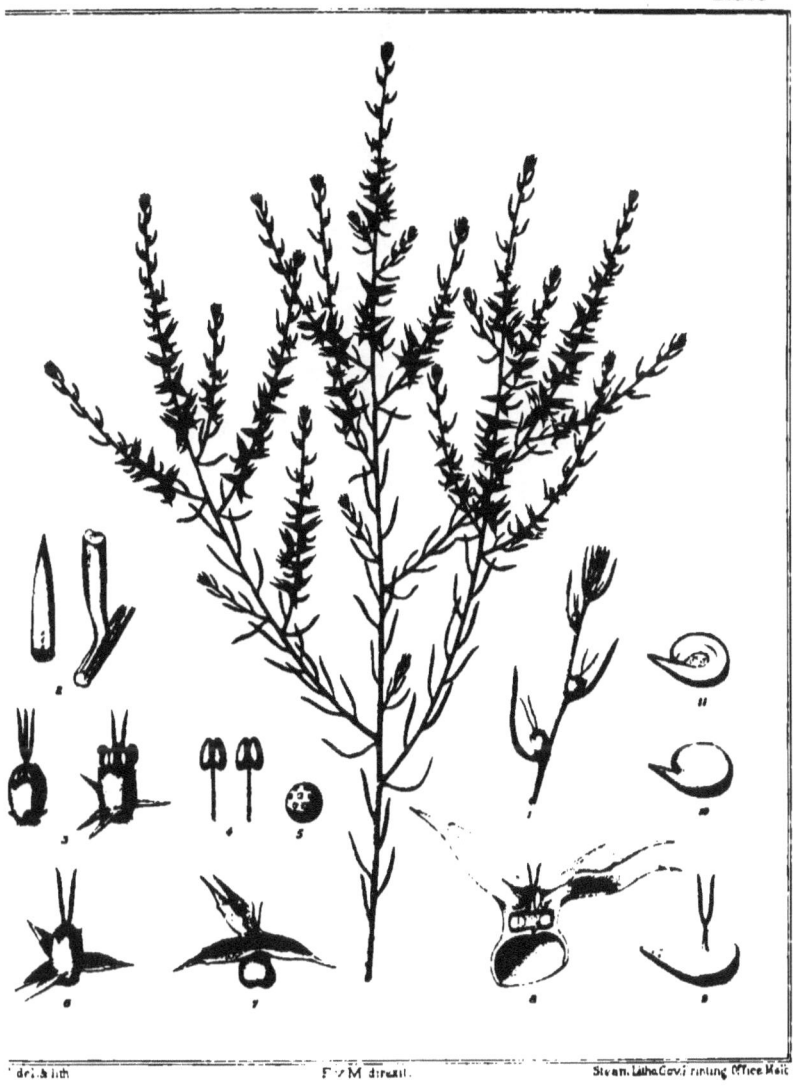

Bassia tridens *FvM*

BASSIA ASTROCARPA.

F. v. M., fragmenta phytographiae Australiae xii, 12 (1882).

PLATE LXV.

1, leaves.
2, a flower.
3, front- and back-view of a stamen.
4, pollen-grain.
5, three fruits with calyx, style and stigmas.
6, vertical section of a fruit with calyx, style and stigmas.
7, a fruit, the calyx removed.
8, a seed.
9, longitudinal section of a seed.

All enlarged, but to various extent.

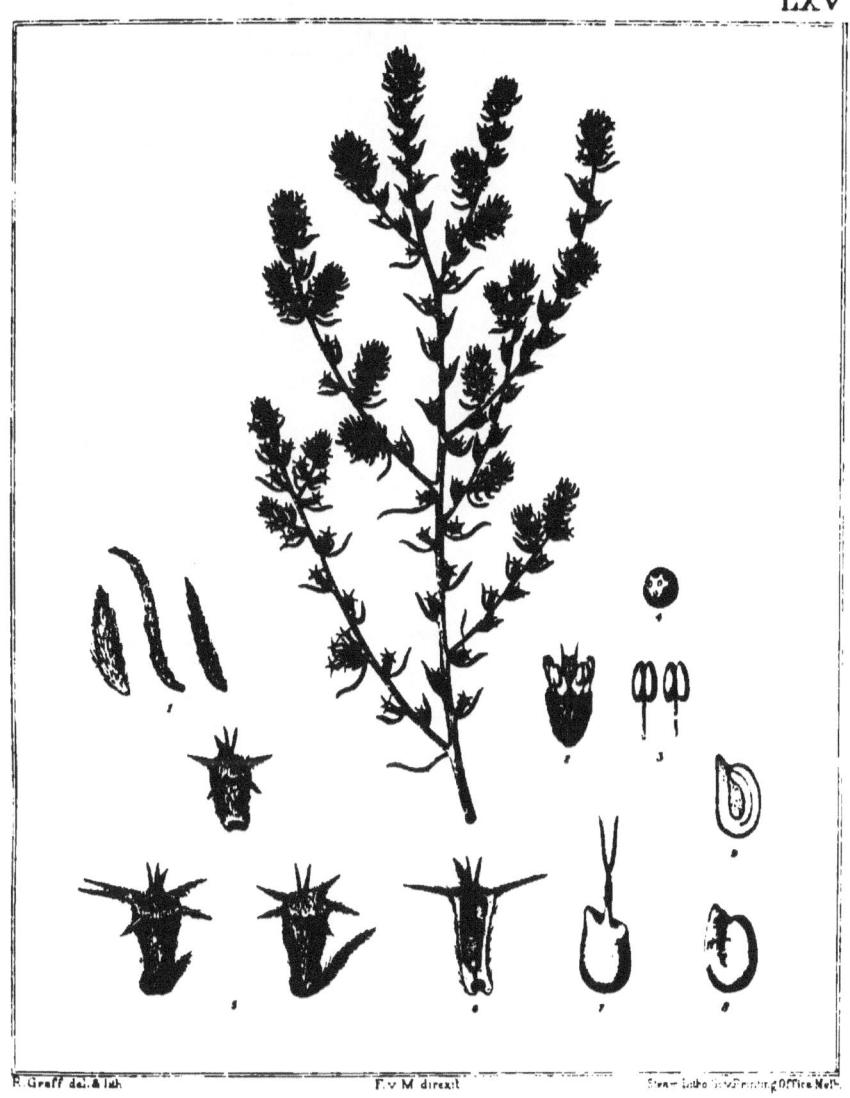

Bassia astrocarpa *FvM*

BASSIA GLABRA.

F. v. M., Census of Australian Plants 30 (1882).

PLATE LXVI.

1 and 2, branchlets with leaves and fruits.

3, a flower.

4, front- and back-view of a stamen.

5, pollen-grain.

6, a fruit with calyx, style and stigmas.

7, vertical section of a fruit with calyx, style and stigmas.

8, a fruit, the calyx removed, the style and stigmas remaining.

9, two seeds.

10, longitudinal section of a seed.

All enlarged, but to various extent.

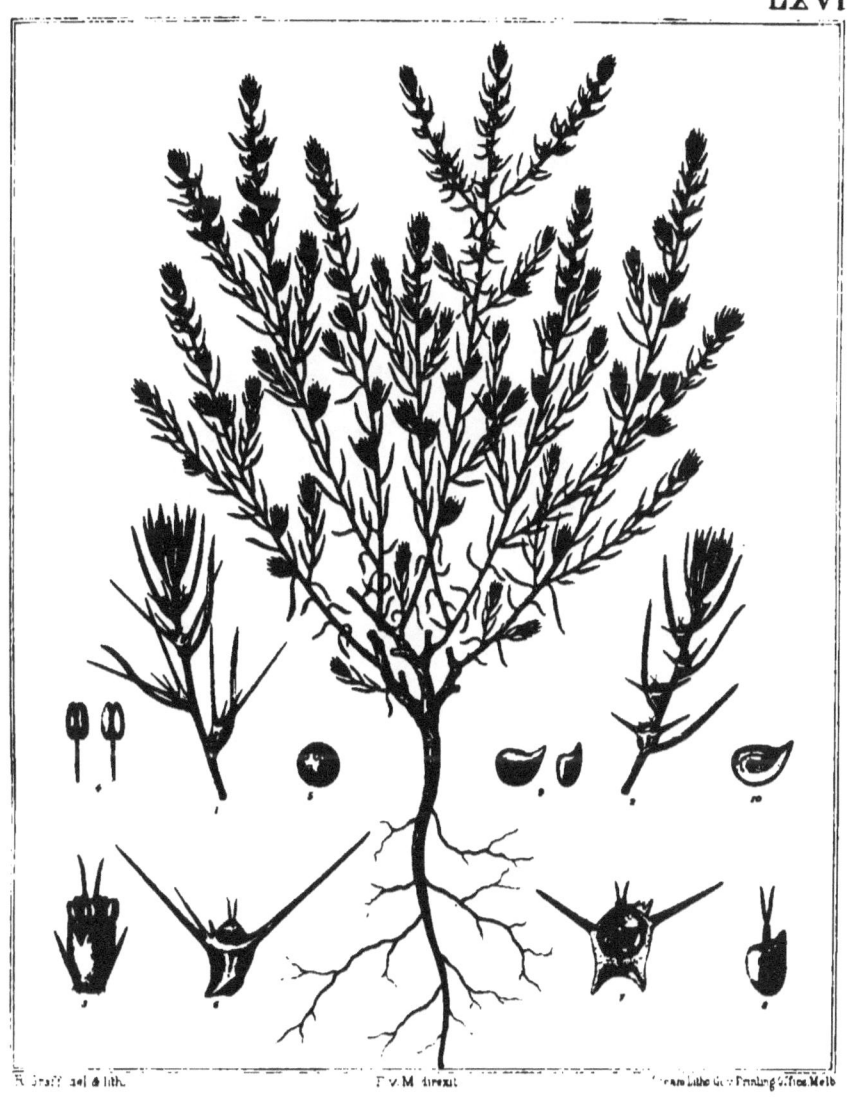

Bassia glabra F.vM.

BASSIA BREVICUSPIS.

F. v. M., Census of Australian Plants 30 (1882).

PLATE LXVII.

1 and 2, portions of two branchlets with flowers and fruits.

3, portions of leaves.

4, flowers in different stages of development.

5, front- and back-view of a stamen.

6, pollen-grain.

7, four different fruits with their calyces, styles and stigmas.

8, vertical section of a fruit with calyx, style and stigmas.

9, a fruit, the calyx removed, the style and stigmas remaining.

10, a seed.

11, longitudinal section of a seed.

All enlarged, but to various extent.

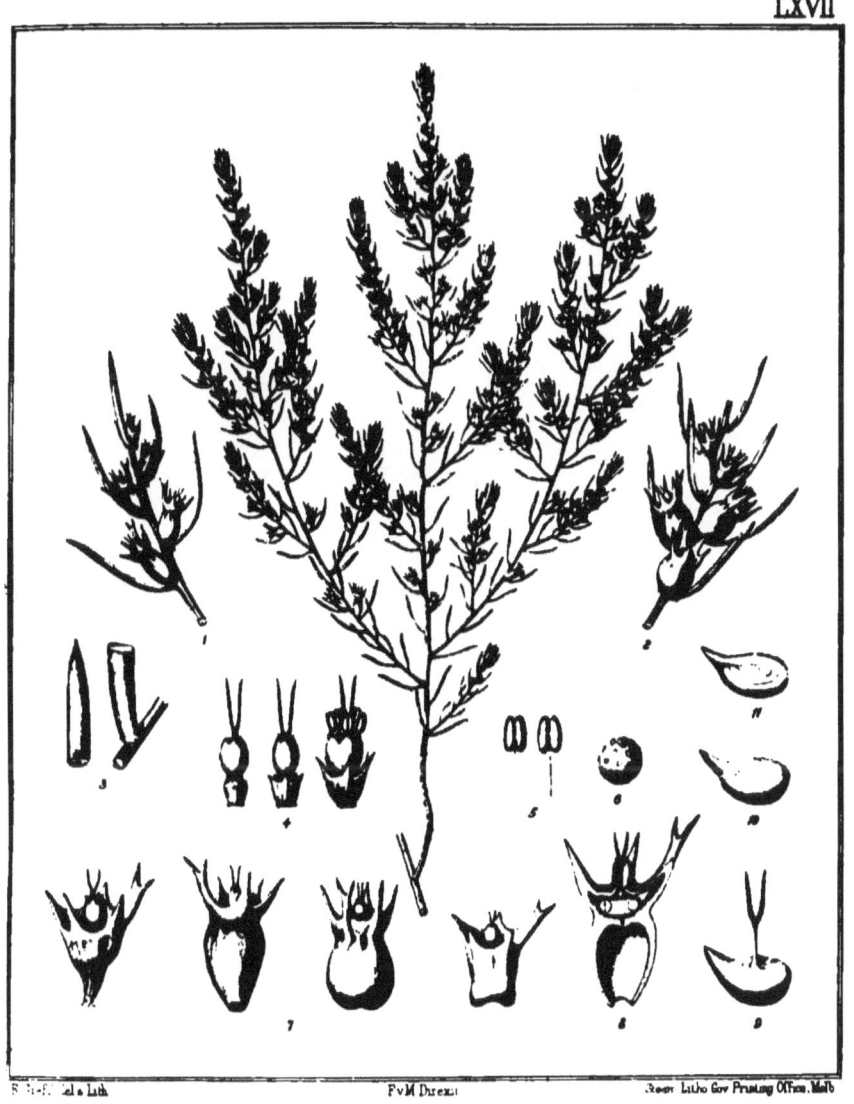

Bassia breviouspis *FvM*

BASSIA STELLIGERA.

Maireana stelligera, F. v. M., fragmenta phytographiae Australiae i, 139 (1859).

PLATE LXVIII.

1, portion of a branchlet with flower and fruits.

2, a flower.

3, front- and back-view of a stamen.

4, pollen-grain.

5, several fruits with calyx, style and stigmas.

6, vertical section of two fruits with calyx, style and stigmas.

7, horizontal section of a fruit with calyx.

8, a fruit, the calyx removed, the style and stigmas remaining.

9, a seed.

All enlarged, but to various extent.

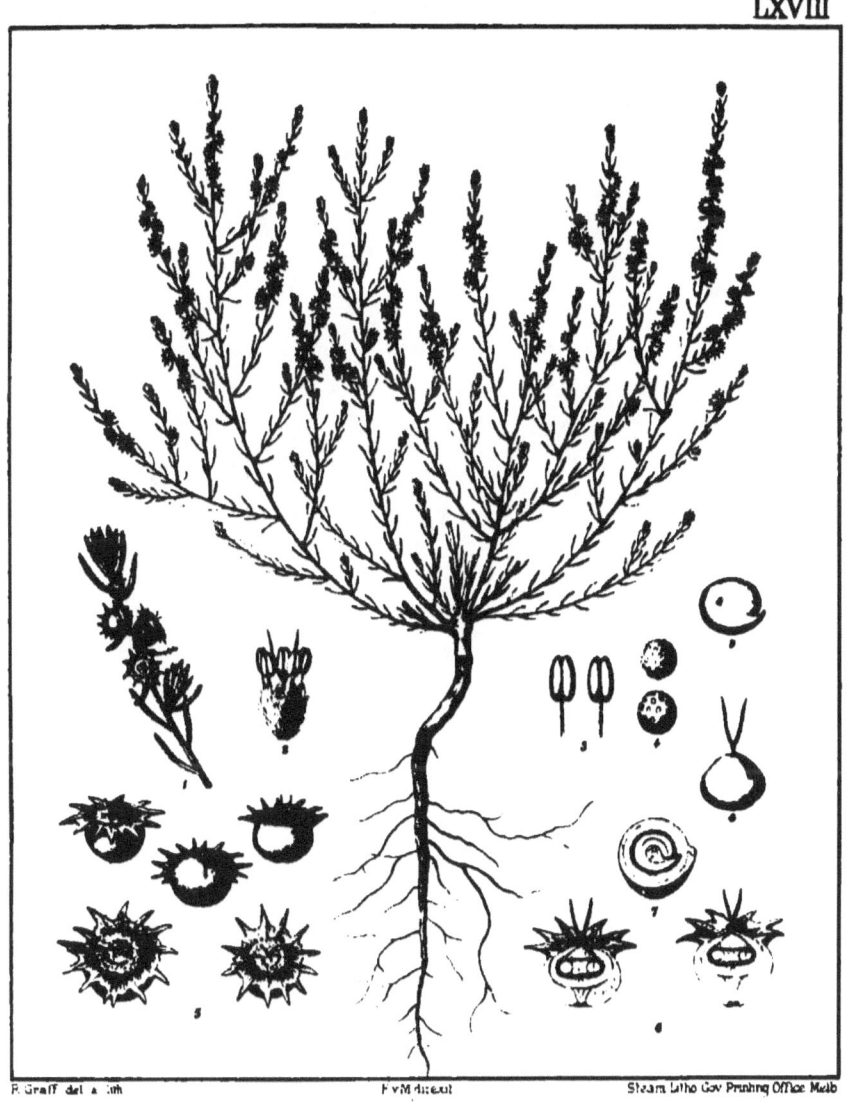

Bassia stelligera FvM

BASSIA ECHINOPSILA.

F. v. M., Census of Australian Plants 30 (1882).

PLATE LXIX.

1, portions of leaves.

2, flowers in different stages of development.

3, a flower, portion of the calyx removed.

4, front- and back-view of a stamen.

5, pollen-grain.

6, fruits with calyx, style and stigmas.

7, vertical section of three fruits with calyx.

8, three fruits, the calyx removed, the stigmas remaining.

9, two seeds.

10, vertical section of two seeds.

11, longitudinal section of a seed.

All enlarged, but to various extent.

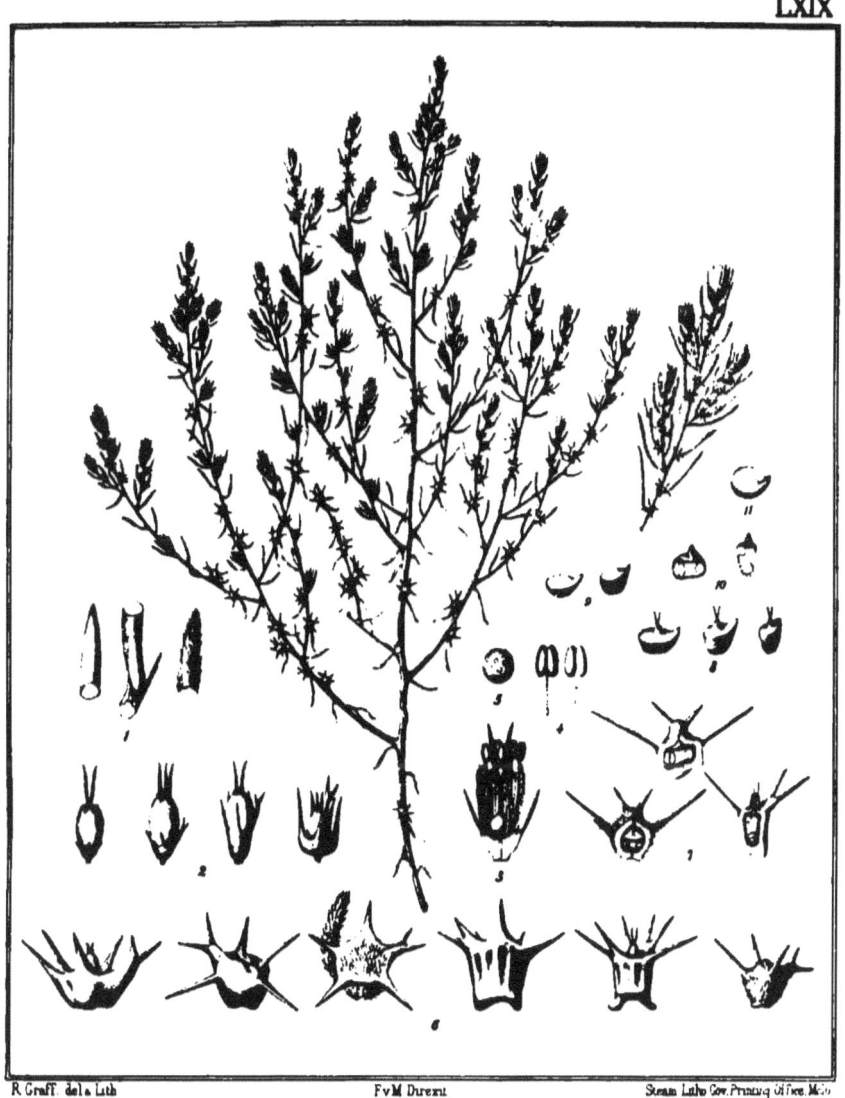

Bassia echinopsila F.v.M.

Bassia Luehmanni.

F. v. M. in Victorian Naturalist vii, 47 (1890).

PLATE LXX.

1, two leaves.

2, a flower.

3, front- and back-view of a stamen.

4, pollen-grain.

5, different fruits with their calyces.

6, vertical section of a fruit with its calyx.

7, a fruit, the calyx removed, the style and stigmas remaining.

8, a seed.

9, longitudinal section of a seed.

All enlarged, but to various extent.

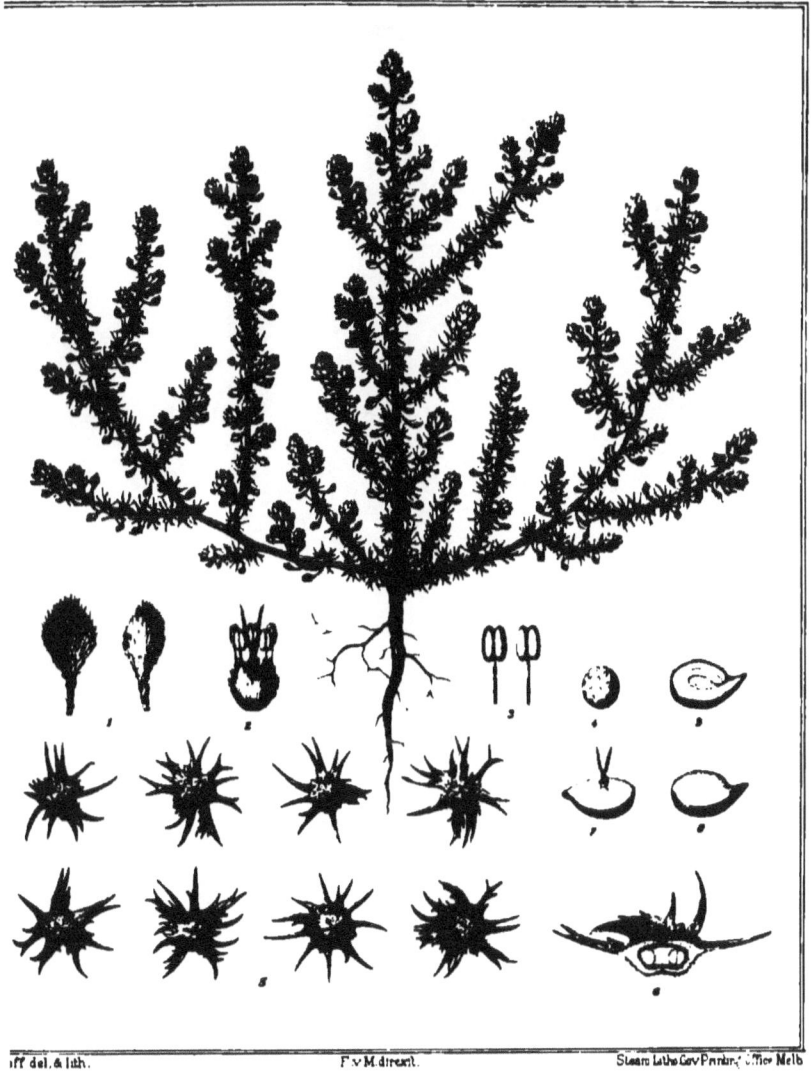

Bassia Luehmanni *FvM*

ICONOGRAPHY

OF

AUSTRALIAN SALSOLACEOUS PLANTS

BY

BARON FERD. VON MUELLER, K.C.M.G., M. & PH.D., F.R.S.,

GOVERNMENT BOTANIST OF THE COLONY OF VICTORIA.

EIGHTH DECADE.

By Authority:
ROBT. S. BRAIN, GOVERNMENT PRINTER, MELBOURNE.
1891.

Bassia Tatei.

F. v. M. in Victorian Naturalist vii., 66 (1890).

PLATE LXXI.

1, portion of a branchlet with fruits.
2, portions of leaves.
3, a flower.
4, front- and back-view of a stamen.
5, pollen-grain.
6 and 7, fruit-bearing calyces.
8, vertical section of a fruit-bearing calyx.
9, a fruit, the calyx removed.
10, a seed.
11, longitudinal section of a seed.

All enlarged, but to various extent.

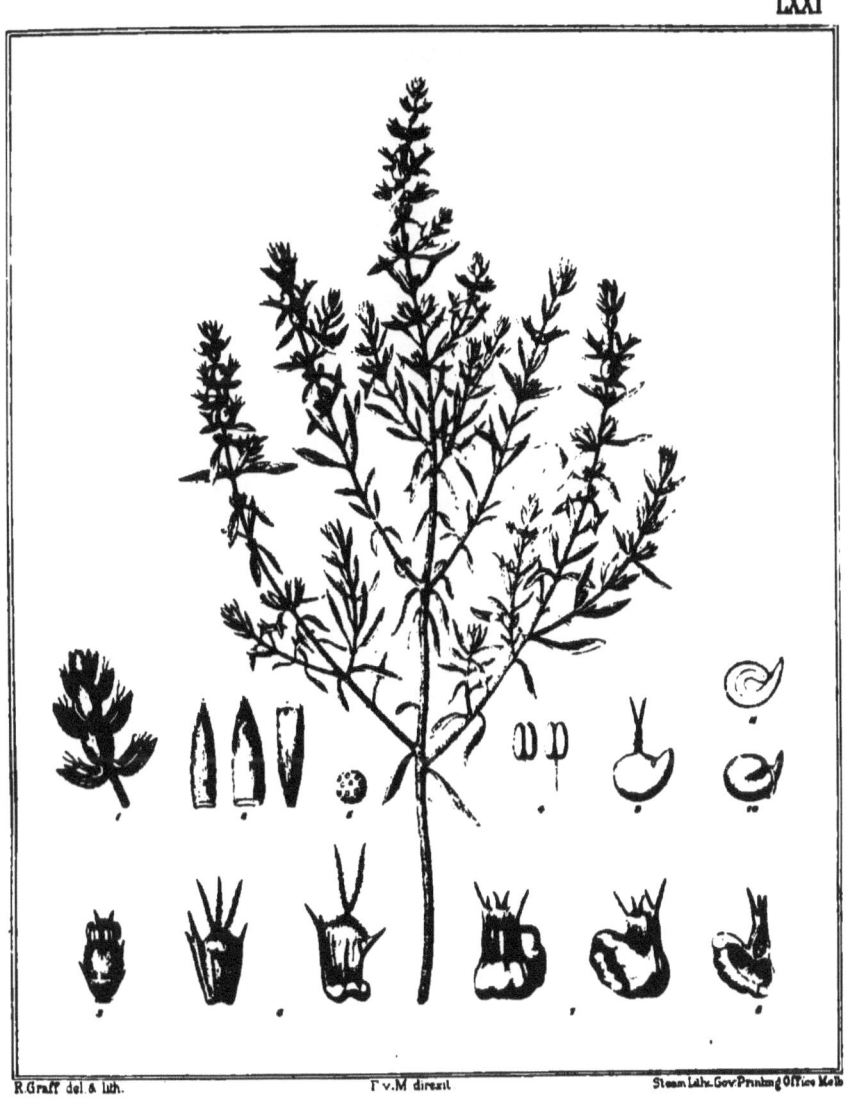

Bassia Tatei F.v.M.

Bassia Birchii.

F. v. M., Census of Australian Plants 30 (1882).

PLATE LXXII.

1, leaves.

2, an unexpanded flower.

3, an expanded flower.

4, front- and back-view of a stamen

5, pollen-grain.

6, fruit-bearing calyces.

7, vertical section of a fruit-bearing calyx.

8, a fruit, the calyx removed.

9, a seed.

10, longitudinal section of a seed.

All enlarged, but to various extent.

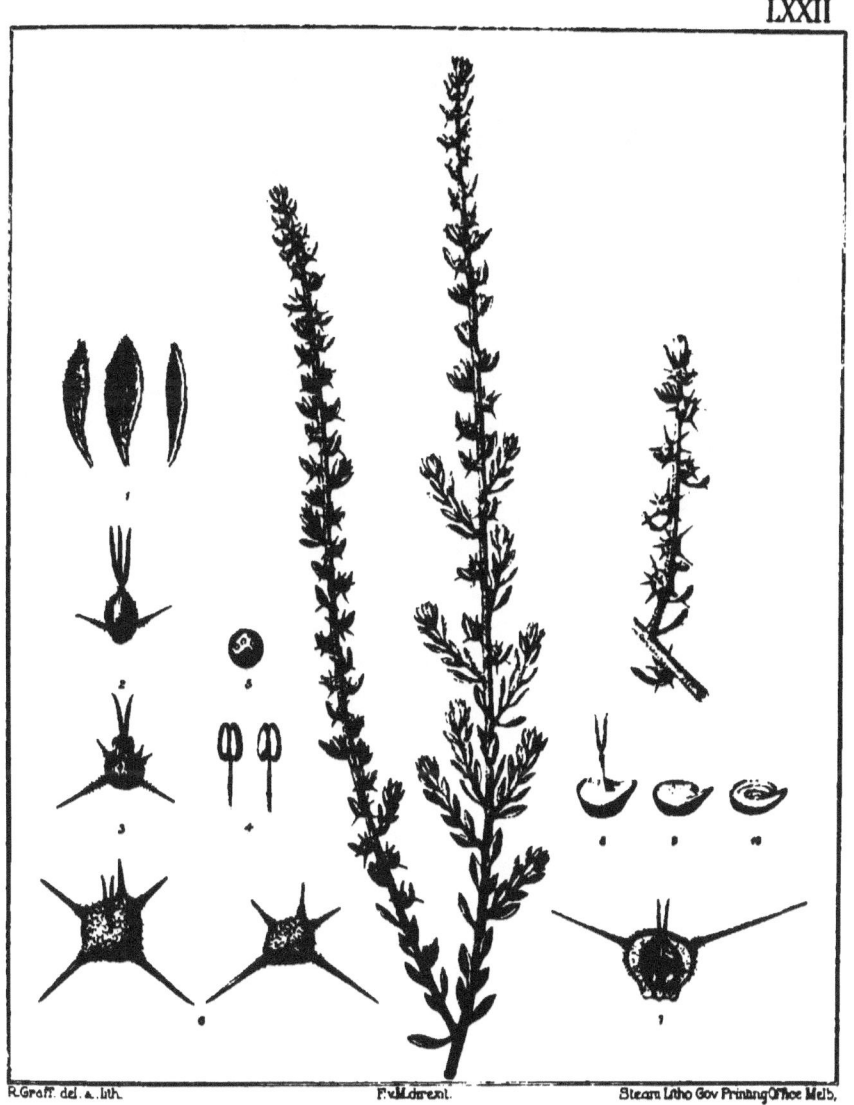

Bassia Birchii *F.v.M.*

BASSIA BICUSPIS.

F. v. M., Census of Australian Plants 30 (1882).

PLATE LXXIII.

1, portions of leaves.
2, an unexpanded flower.
3, expanded flowers.
4, front- and back-view of a stamen.
5, pollen-grain.
6, longitudinal section of a fruit-bearing calyx.
7, a fruit, the calyx removed.
8, a seed.
9, transverse section of a seed.
10, longitudinal section of a seed.

All enlarged, but to various extent.

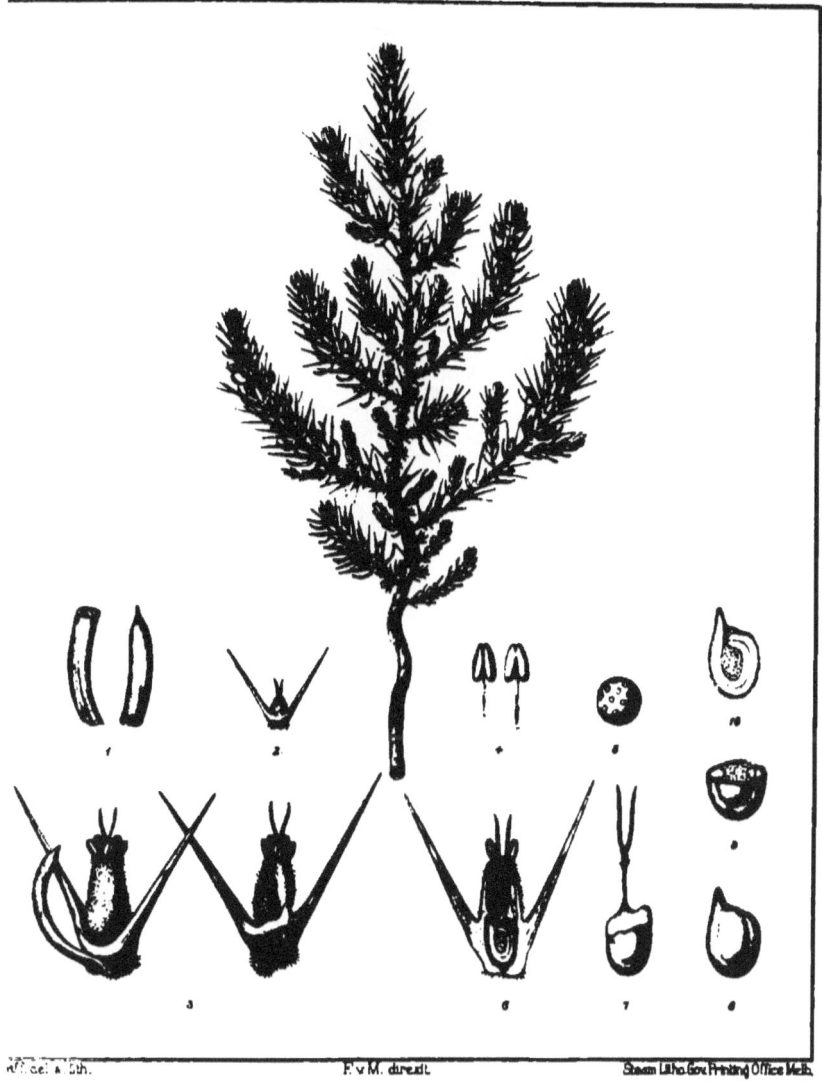

Bassia bicuspis *F.v.M.*

Bassia longicuspis.

F. v. M., inedited.

PLATE LXXIV.

1, portions of leaves.
2, a flower.
3, front- and back-view of a stamen.
4, pollen-grain.
5, fruit-bearing calyces.
6, vertical section of a fruit-bearing calyx.
7, fruits, the calyces removed.
8, a seed.
9, transverse section of a seed.
10, longitudinal section of a seed.

All enlarged, but to various extent.

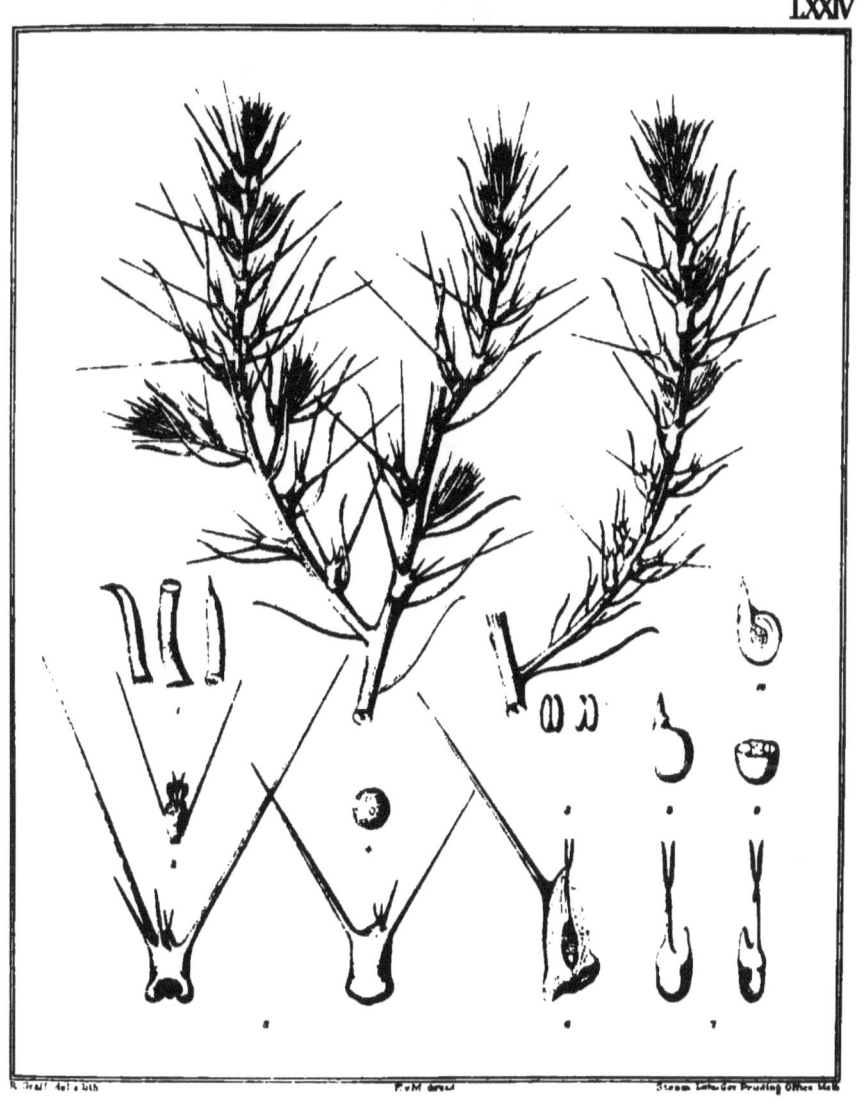

Bassia longicuspis *F.v.M.*

Bassia Forrestiana.

F. v. M., fragmenta phytographiae Australiae xii., 12 (1882).

PLATE LXXV.

1, portions of leaves.
2, a flower.
3, front- and back-view of a stamen.
4, pollen-grain.
5, a fruit-bearing calyx.
6, vertical section of a fruit-bearing calyx.
7, a fruit, the calyx removed.
8, a seed.
9, transverse section of a seed.
10, longitudinal section of a seed.

All enlarged, but to various extent.

Bassia Forrestiana *F.v.M.*

BASSIA QUINQUECUSPIS.

F. v. M., Census of Australian Plants 30 (1882).

PLATE LXXVI.

1, portions of leaves.
2, flowers.
3, front- and back-view of a stamen.
4, pollen-grain.
5, a fruit-bearing calyx.
6, vertical section of a fruit-bearing calyx.
7, a fruit, the calyx removed.
8, a seed.
9, transverse section of a seed.

All enlarged, but to various extent.

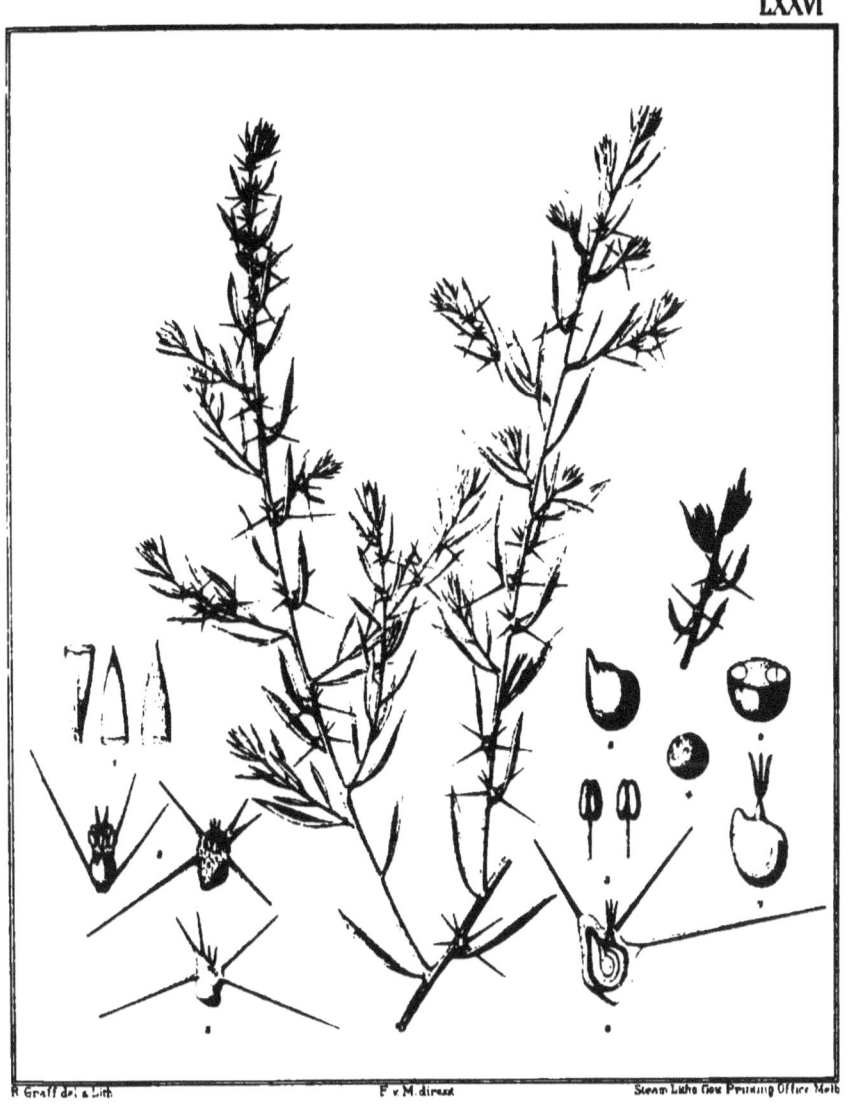

Bassia quinquecuspis F.v.M.

BASSIA DIVARICATA.

F. v. M., Census of Australian Plants 30 (1882).

PLATE LXXVII.

1, portions of leaves.
2, unexpanded flowers.
3, an expanded flower.
4, front- and back-view of a stamen.
5, pollen-grain.
6, fruit-bearing calyces.
7, vertical section of a fruit-bearing calyx.
8, fruits, the calyces removed.
9, a seed.
10, transverse section of a seed.
11, longitudinal section of a seed.

All enlarged, but to various extent.

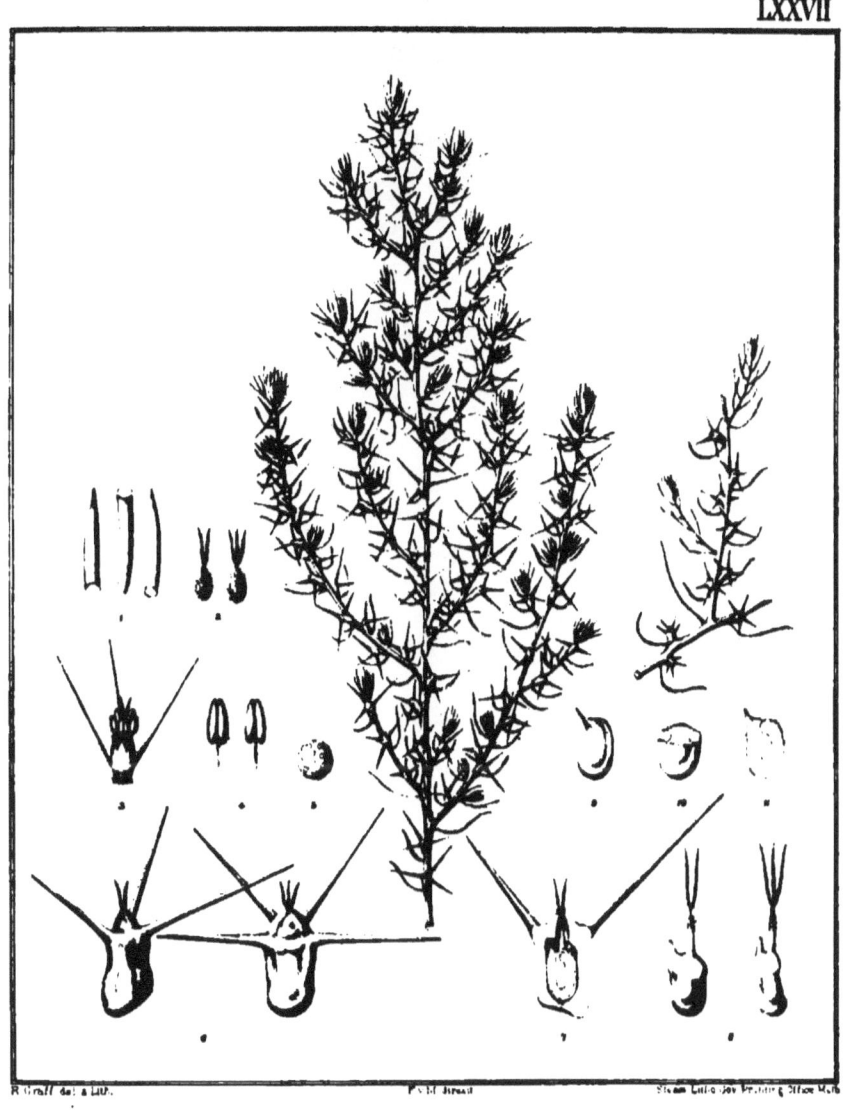

Bassia divaricata F.v.M.

BASSIA DIACANTHA.

F. v. M., Census of Australian Plants 30 (1882)

PLATE LXXVIII.

1, portions of leaves.
2, a flower.
3, front- and back-view of a stamen.
4, pollen-grain.
5, fruit-bearing calyces.
6, vertical sections of two fruit-bearing calyces.
7, a fruit, the calyx removed.
8, a seed.
9, transverse section of a seed.
10, longitudinal section of a seed.

All enlarged, but to various extent.

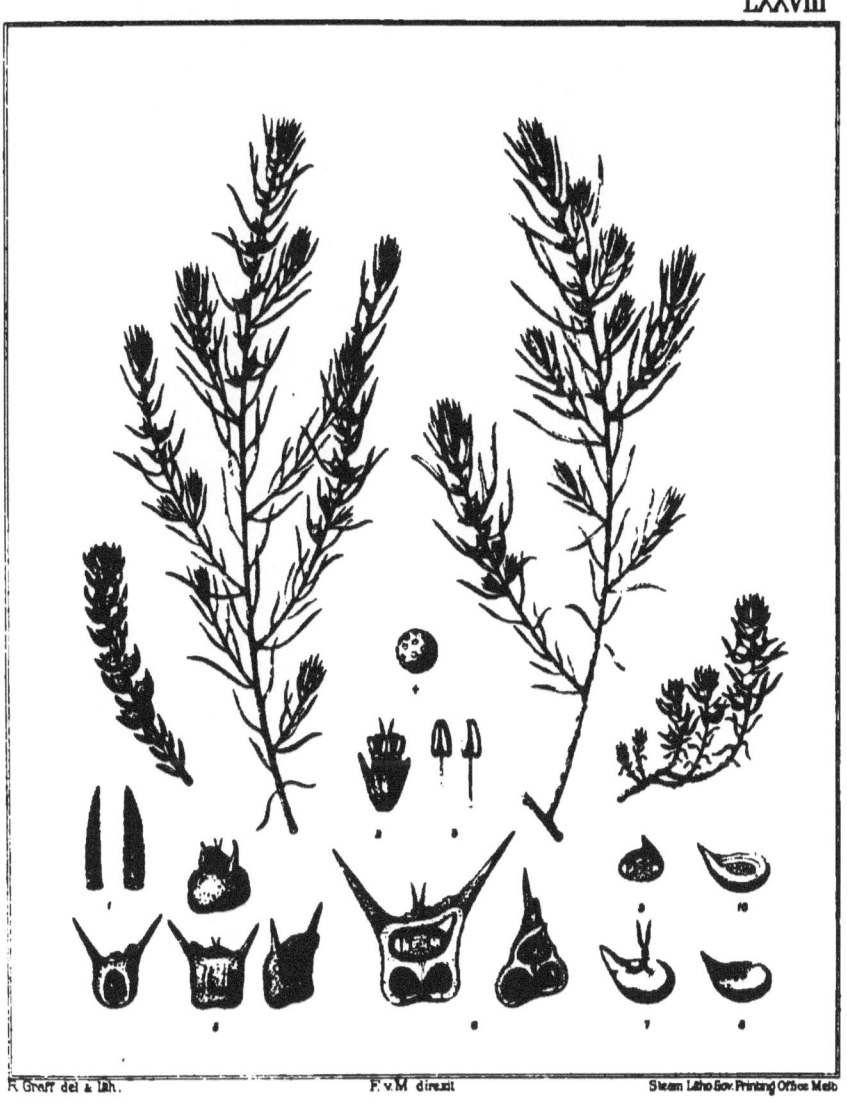

Bassia diacantha F.v.M.

BASSIA BICORNIS.

F. v. M., Census of Australian Plants 30 (1882).

PLATE LXXIX.

1, portions of leaves.
2, a flower.
3, front- and back-view of a stamen.
4, pollen-grain.
5, vertical section of a fruit-bearing calyx.
6, a fruit, the calyx removed.
7, a seed.
8, transverse section of a seed.
9, longitudinal section of a seed.

All enlarged, but to various extent.

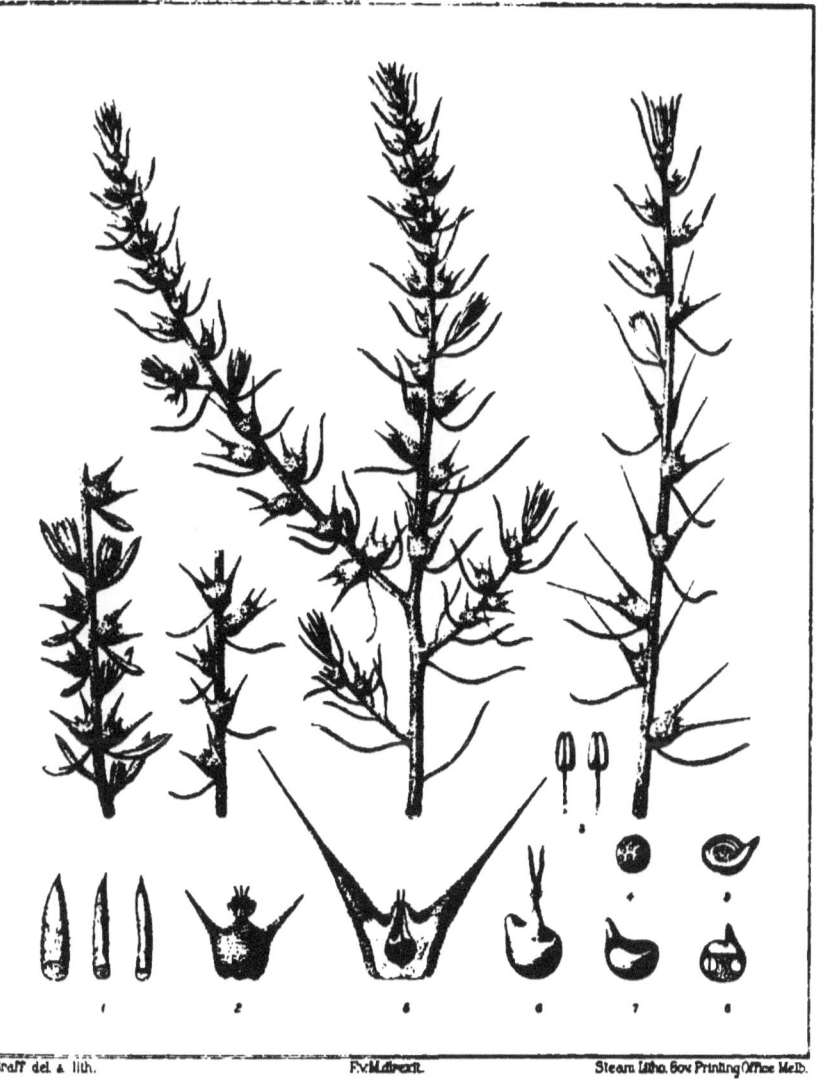

Bassia bicornis F.v.M.

BASSIA LANICUSPIS.

F. v. M., Census of Australian Plants 30 (1882).

PLATE LXXX.

1, portions of leaves.
2, flowers.
3, front- and back-view of a stamen.
4, pollen-grain.
5, fruit-bearing calyces.
6, vertical section of a fruit-bearing calyx.
7, a fruit, the calyx removed.
8, a seed.
9, transverse section of a seed.
10, longitudinal section of a seed.

All enlarged, but to various extent.

Bassia lanicuspis *F.vM.*

ICONOGRAPHY

OF

AUSTRALIAN SALSOLACEOUS PLANTS,

BY

BARON FERD. VON MUELLER, K.C.M.G., M. & PH.D., F.R.S.,

GOVERNMENT BOTANIST OF THE COLONY OF VICTORIA.

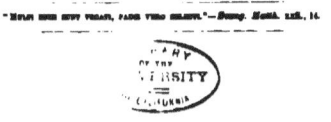

NINTH DECADE.

By Authority:
ROBT. S. BRAIN, GOVERNMENT PRINTER, MELBOURNE.
1891.

Bassia Eurotioides.

F. v. M., Census of Australian Plants 30 (1882).

PLATE LXXXI.

1, a flower.
2, front- and back-view of a stamen.
3, pollen-grain.
4, a fruit-bearing calyx.
5, vertical section of a fruit-bearing calyx.
6, a fruit, the calyx removed.
7, a seed.
8, longitudinal section of a seed.

All enlarged, but to various extent.

Bassia eurotioides *F.v.M*

BASSIA SCLEROLAENOIDES.

Gras in Bulletin de la Société botanique de France (1864), implied.

PLATE LXXXII.

1, portion of a branchlet.
2, portions of a leaf.
3, unexpanded flowers.
4, expanded flower.
5, front- and back-view of a stamen.
6, pollen-grain.
7 and 8, fruit-bearing calyces.
9, vertical section of a fruit-bearing calyx.
10, a fruit, the calyx removed.
11, a seed.
12, transverse section of a seed.
13, longitudinal section of a seed.

All enlarged, but to various extent.

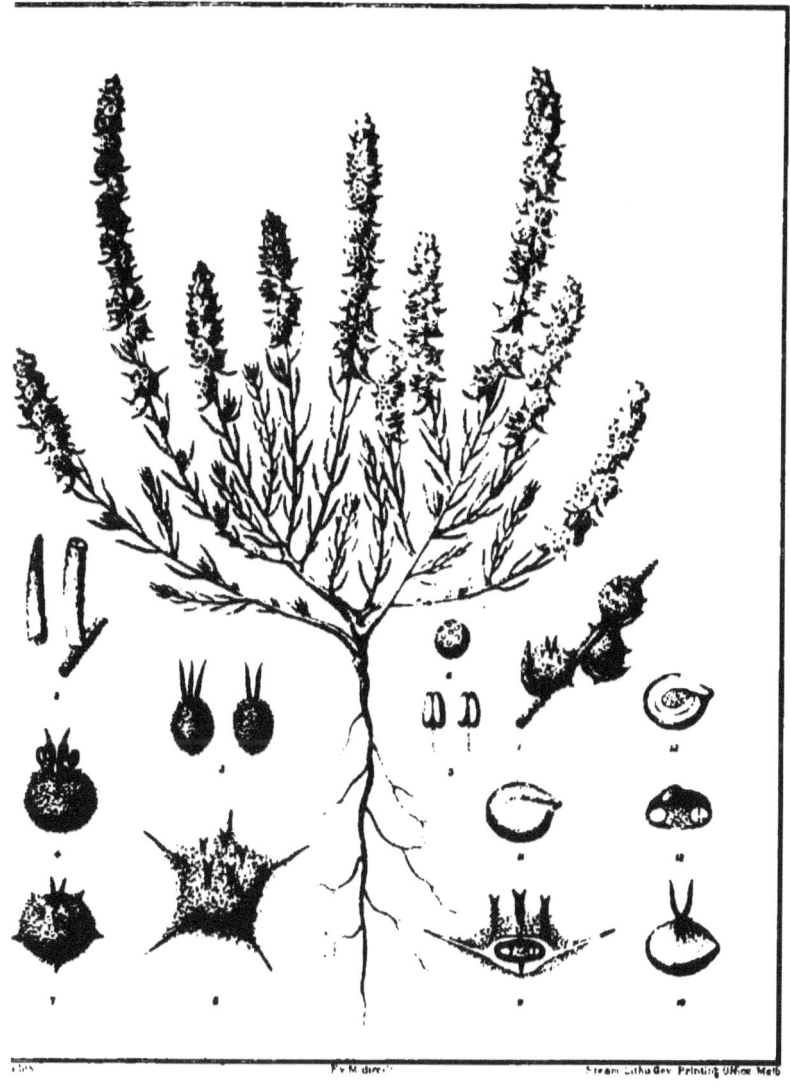

Bassia sclerolaenoides *Gras.*

BASSIA CARNOSA.

F. v. M., Census of Australian Plants 30 (1882).

PLATE LXXXIII.

1, leaves.
2, portions of leaves.
3, flowers.
4, front- and back-view of a stamen.
5, pollen-grain.
6, top- and bottom-view of a fruit-bearing calyx.
7, vertical section of a fruit-bearing calyx.
8, a fruit, the calyx removed.
9, a seed.
10, transverse section of a seed.
11, longitudinal section of a seed.

All enlarged, but to various extent.

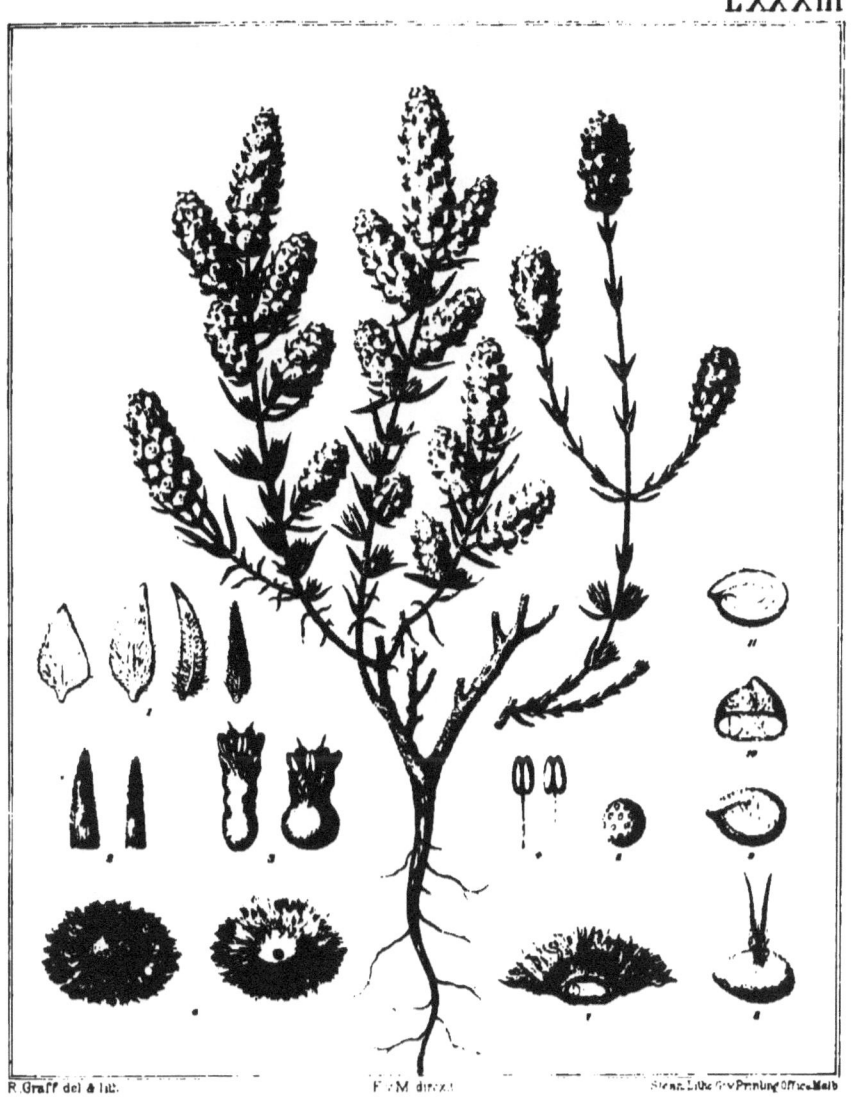
Bassia carnosa F.v.M

Bassia enchylaenoides.

F. v. M., Census of Australian Plants 30 (1882).

PLATE LXXXIV.

1, portion of a branchlet.

2 and 3, leaves, portions and young.

4, a flower.

5, front- and back-view of a stamen.

6, pollen-grain.

7, a flower, part of the calyx removed.

8 and 9, fruit-bearing calyces, top- and bottom-view.

10, a fruit, the calyx removed.

11, a seed.

12, transverse section of a seed.

13, longitudinal section of a seed.

All enlarged, but to various extent.

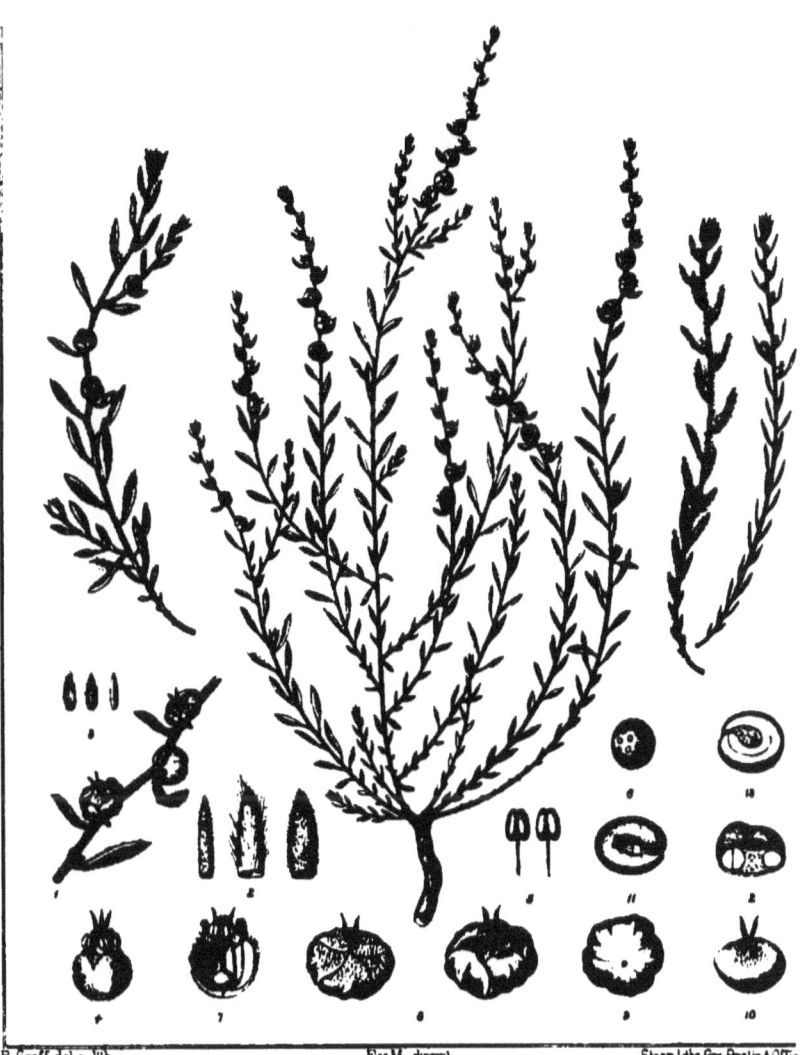

Bassia enchylaenoides F.v.M.

ENCHYLAENA TOMENTOSA.

R. Brown, prodromus florae Novae Hollandiae 408 (1810).

PLATE LXXXV.

1, portions of leaves.
2, flowers.
3, front- and back-view of a stamen.
4, pollen-grain.
5, fruit-bearing calyces.
6, vertical section of a fruit-bearing calyx.
7, a fruit, the calyx removed.
8, a fruit, seen from beneath.
9, a seed.
10, longitudinal section of a seed.

All enlarged, but to various extent.

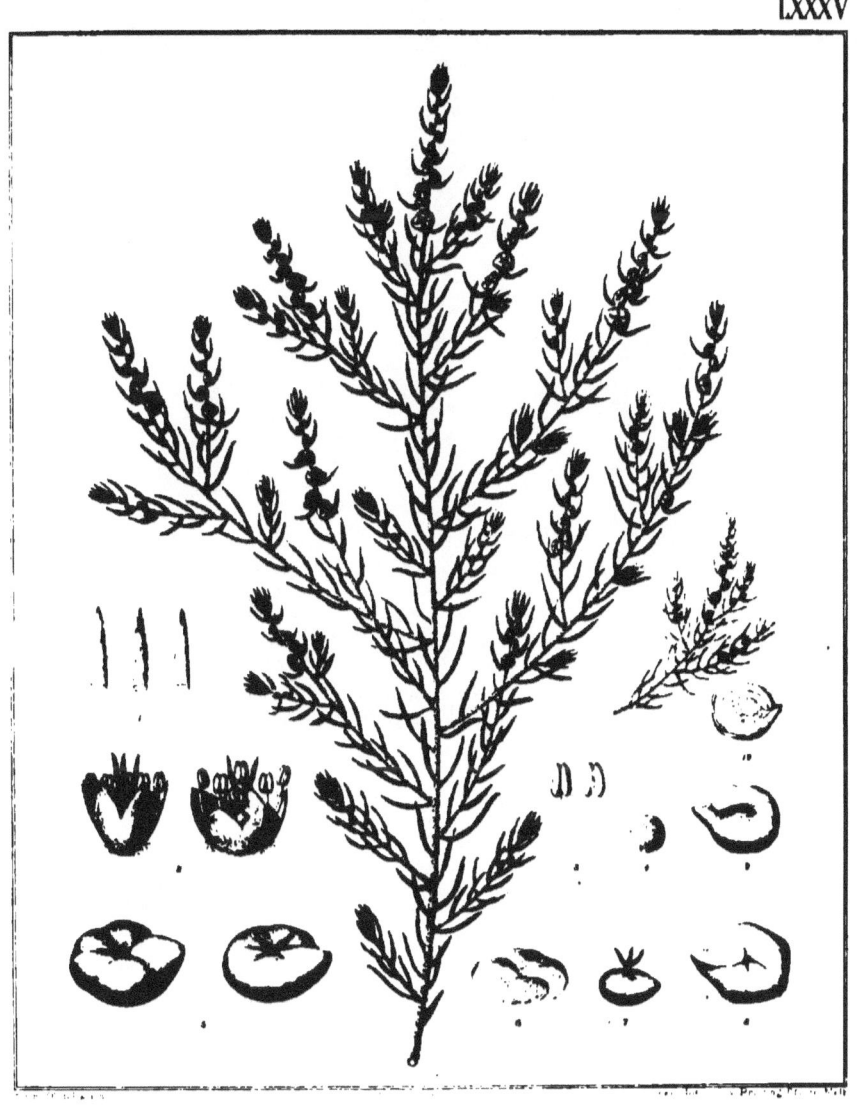

Enchylaena tomentosa R. BROWN.

THRELKELDIA DIFFUSA.

R. Brown, prodromus florae Novae Hollandiae 410 (1810).

PLATE LXXXVI.

1, portions of leaves.
2, an unexpanded flower.
3, an expanded flower.
4, front- and back-view of a stamen.
5, pollen-grain.
6, 7, and 8, fruit-bearing calyces.
9, vertical section of a fruit-bearing calyx.
10, a fruit, the calyx removed.
11, a seed.
12, longitudinal section of a seed.

All enlarged, but to various extent.

THRELKELDIA PROCERIFLORA.

F. v. M., fragmenta phytographiae Australiae viii., 38 (1873).

PLATE LXXXVII.

1, portions of a leaf.
2, an unexpanded flower.
3, an expanded flower.
4, front- and back-view of a stamen.
5, pollen-grain.
6 and 7, fruit-bearing calyces.
8, vertical section of a fruit-bearing calyx.
9, a fruit, the calyx removed.
10, a seed.
11, transverse section of a seed.
12, longitudinal section of a seed.

All enlarged, but to various extent.

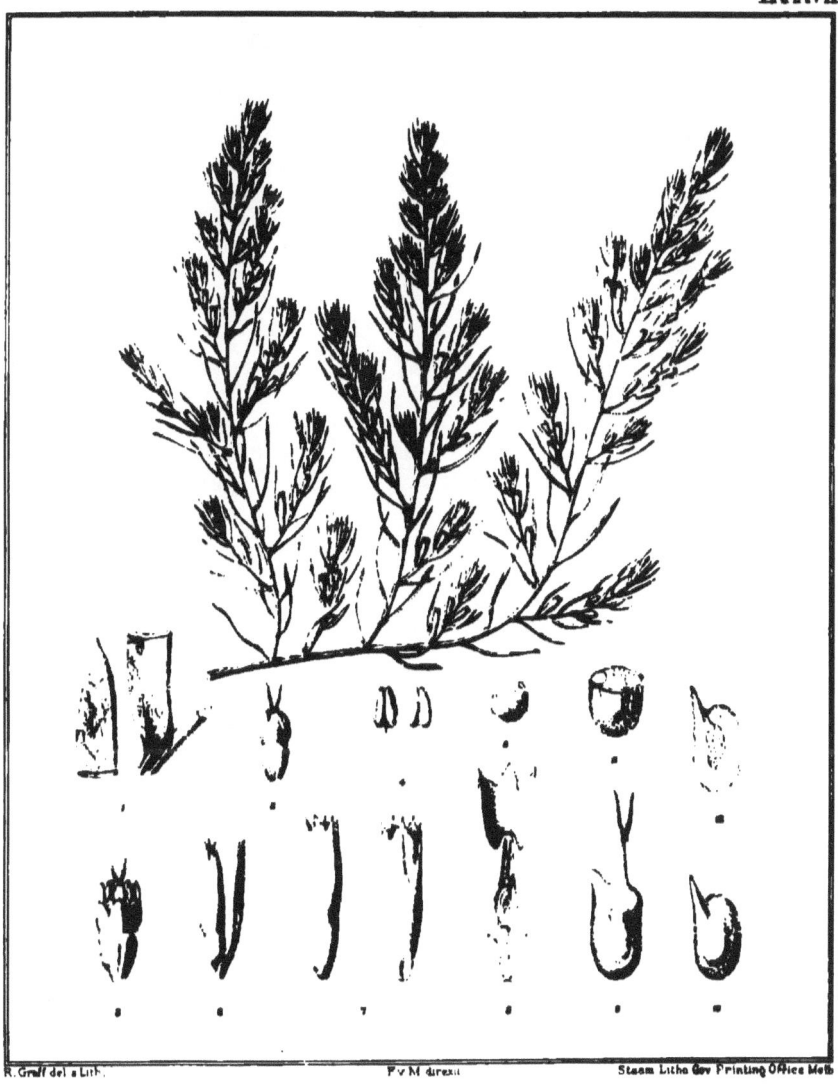

LXXXVII

OSTEOCARPUM SALSUGINOSUM.

F. v. M., Second General Report 15 (1855).

PLATE LXXXVIII.

1, portions of a leaf.
2, a flower.
3, front- and back-view of a stamen.
4, pollen-grain.
5 and 6, fruit-bearing calyces.
7, vertical section of a fruit-bearing calyx.
8, a fruit, the calyx removed.
9, a seed.
10, longitudinal section of a seed.

All enlarged, but to various extent.

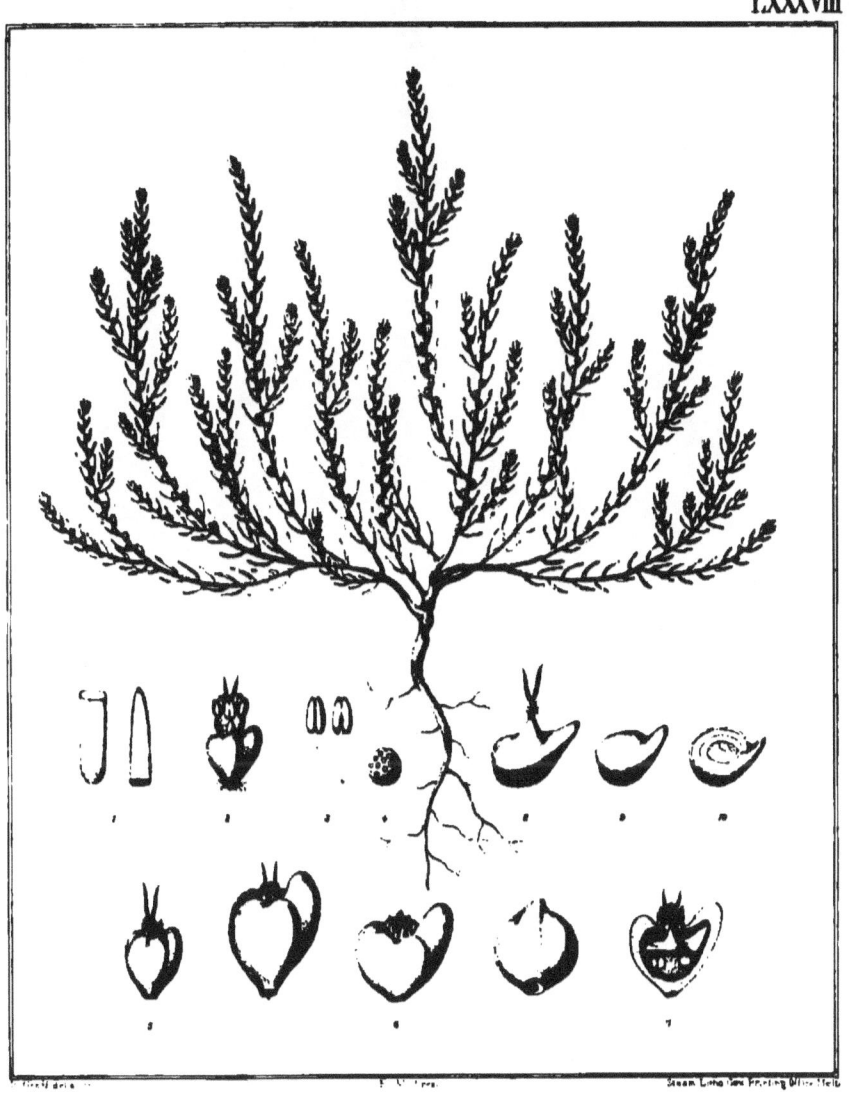

Osteocarpum salsuginosum *F.v.M.*

SUAEDA MARITIMA.

Dumortier, Florula Belgica 22 (1827).

PLATE LXXXIX.

1, portions of a leaf.
2, portion of a branchlet with leaves and flowers.
3, a flower seen from beneath.
4, side-view of a flower.
5, front- and back-view of a stamen.
6, pollen-grain.
7, a fruit-bearing calyx.
8, vertical section of a fruit-bearing calyx.
9, a fruit, the calyx removed.
10, a seed.
11, transverse section of a seed.
12, longitudinal section of a seed.
13, embryo.

All enlarged, but to various extent.

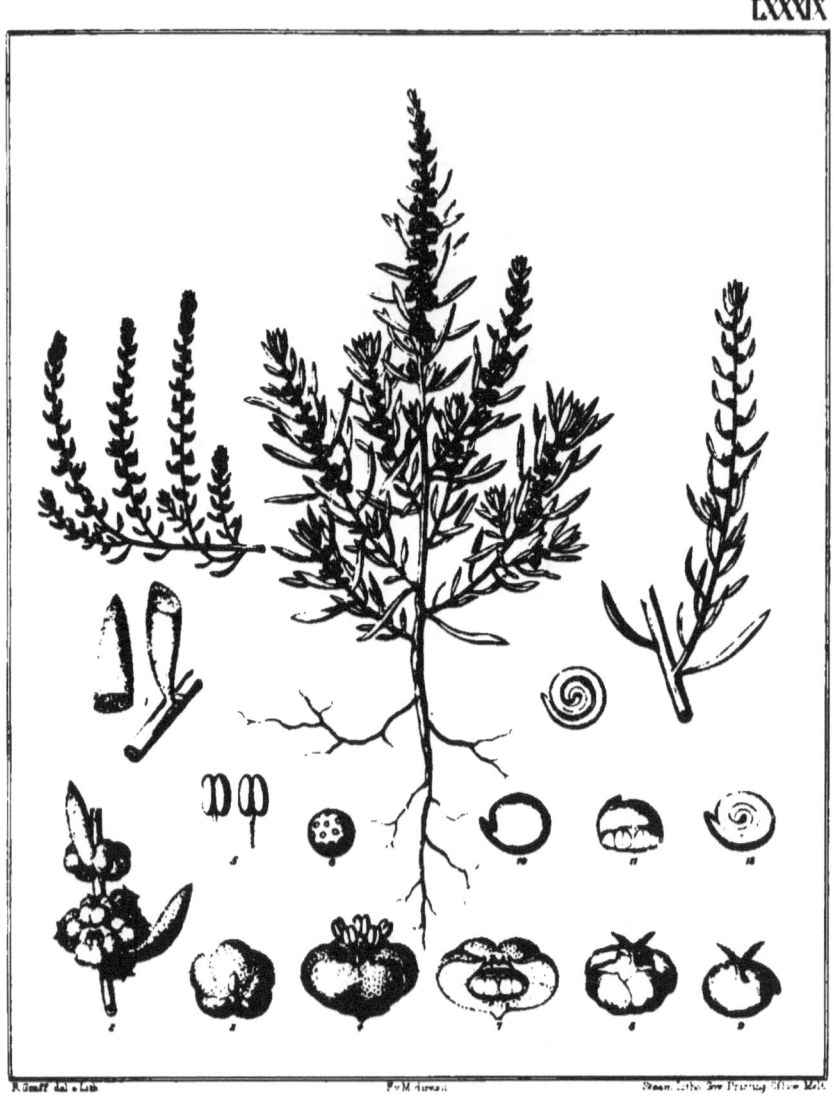

Suaeda maritima *Dumortier*

SALSOLA KALI.

Linné, Species Plantarum 222 (1753).

PLATE XC.

1, portion of a branchlet with a flower, floral leaf and bracteoles.

2, a flower, the calyx removed.

3, back- and front-view of a stamen.

4, pollen-grain.

5, immature fruit-bearing calyces.

6, a fruit-bearing calyx with floral leaf and bracteoles.

7, fruit-bearing calyces.

8, vertical section of a fruit-bearing calyx.

9, fruits, the calyces removed.

10, embryos, extracted.

11, longitudinal section of a seed.

All enlarged, but to various extent.

Salsola Kali *Linné.*

www.ingramcontent.com/pod-product-compliance
Lightning Source LLC
Chambersburg PA
CBHW031424230426
43668CB00007B/422